On the Logic of Ordinary Conditionals

SUNY Series in Logic and Language

John T. Kearns, Editor

On the Logic of Ordinary Conditionals

Robert N. McLaughlin

STATE UNIVERSITY OF NEW YORK PRESS

Published by
State University of New York Press, Albany

Printed in the United States of America

For information, address State University of New York Press,
State University Plaza, Albany, N.Y., 12246

Library of Congress Cataloging in Publication Data

McLaughlin, Robert N., 1928–
On the logic of ordinary conditionals / Robert N. McLaughlin.
p. cm. — (SUNY Series in logic and language)
Includes bibliographical references.
ISBN 0-7914-0293-2. — ISBN 0-7914-0294-0
1. Conditionals (Logic) I. Title. II. Series.
BC199.C56M35 1990
160–dc20 89-21611
CIP

10 9 8 7 6 5 4 3 2 1

Contents

To Claire

Preface

We use conditional assertions to express the way things are related to each other. The things that we say are conditionally related, may be happenings in the world or linguistic expressions or concepts. This essay is about the logic of a type of conditional that I believe is present at all levels of language use; it is the conditional we employ when, for example, in asserting a statement of the form, 'If A then, . . . ' we wish to say that if A is the case, then this is sufficient to ensure that something else is the case. If A is some happening, the 'something else' may be simply that another event has occurred or will occur. In this case, the conditional is said to be a factual one, and it can be shown to be false if it should turn out that A occurs and the other event does not. But a conditional may say that if A is the case, then what is warranted or assured is the conclusion that something else *ought* to take place, or that the occurrence of something else is desirable; and such a conditional is not made false if, on the happening of A, the thing that ought to happen, or the thing that is desirable, does not occur. Factuality, then, is not a necessary feature of the conditional I will be discussing.

But although there are many types of conditionals that are of the genus in question whose logic would reward exploration, I restrict myself in this essay to the study of three types, which I will name the 'factual', the 'conventional', and the 'conceptual' or 'logical' conditionals. What the factual conditional says is that the fulfillment of the 'if' clause ensures the occurrence of something; the conventional conditional says that if the application of one predicative expression to a thing is accepted, then the acceptance of the application of another predicate to that thing is required by the conventions of a language; and the conceptual or logical conditional, for example, 'Logically necessarily, if A entails B, then A is inconsistent with not B,' says that if the 'if' clause is fulfilled, the denial of the 'then' clause would be absurd, contrary to reason. One might say that the 'if' clause is related to the 'then' clause in the factual conditional as a 'naturally sufficient condition', in the conventional conditional as a 'conventionally sufficient condition' and in the necessary conditional as a 'rationally sufficient condition'. . .

The factual conditional is the main topic of chapters 1, 2, and 3. The characteristics of conventional and logical conditionals are studied in chapters 4 and 6, respectively. My reason for restricting the discussion to these three types of conditionals is that the principal object of the essay is to present a theory of the logical role of conditionality in the language of fact and I believe that although an acceptable elucidation of the logic of factual conditionals must take into consideration aspects of the logic of conventional and logical conditionals, the logic of conditionals having as their subjects such other matters as moral, practical, or aesthetic relations does not form an essential part of such a study.

In saying that the factual conditional 'If . . . then———' asserts that the occurrence of ' . . .' ensures or warrants the occurrence of '———', I do not intend to suggest that the logic of the conditional is to be appraised in terms of the concepts of 'assertibility' or 'rational belief'—in terms of the question whether, given '. . .', '———' is assertible, or that it is rational to believe '———'. The correctness or otherwise of such claims turns as much upon the situation of, the state of knowledge of the speaker, as it does upon the things to which '. . .' and '———' have reference. The factual conditional of this essay concerns the relations present in the situation to which it is addressed, and it is true if the occurrence of what is expressed by the 'if' clause ensures the occurrence of what is expressed by the 'then' clause.

I propose to call the conditional of the type to be examined the 'ordinary' conditional in part to distinguish it from the material conditional of truth functional logic and in part because of its pervasiveness, its importance to all the levels of language I will consider. The name 'ordinary' will be used to differentiate the conditional with which we will be concerned from various other conditionals, such as 'suppositive', 'disposition' and 'even if' conditionals, a conditional I shall call the 'betting conditional' and statements of conditional probability, from all of which the ordinary conditional will be distinguished. As long as one remembers that not all conditionals to be found in natural language may be classified as what I have called 'ordinary', he will not be misled. The system that is offered in these pages to express the logic of the ordinary conditional I shall refer to as OC.

I take the view with regard to statements generally, and ordinary conditionals in particular, that the logic of a statement is determined by the relations it bears to its direct confirmation and disconfirmation (in a sense of these words to be explained in chapter 1). Statements that have similar relations to their confirmations and disconfirmations have similar logics. Factual, conventional, and logical conditionals are similar in this way. The contents of their confirmations and disconfirmations are of different general types and they consequently belong to different levels of language. Factual conditionals are supported by occurrences of the events they are about, conventional ones by accepted uses of expressions, and logical conditionals by our recognition of the limits of intelligibility. But the conditionals I will study all qualify as ordinary conditionals in that the relations they bear to their confirmations are formally similar.

In order to avoid future confusion I mention here two characteristics that an ordinary conditional may but need not have and also two characteristics it never has. Many factual conditionals indicate that the event to which the 'if' clause refers occurs, if it does occur, at a time prior to the event mentioned by the 'then' clause. 'If Smith supports the welfare system in his campaign, he will win the election' is such a conditional. But conventional and logical conditionals are clearly not of this sort, nor are all factual conditionals. A temporal factor is not implied by 'If the rock is from this area then it contains porphyry'. And sometimes the 'if' clause refers to an event that may occur after the event referred to by the 'then' clause: 'If Dave fails his examination, then he did not work hard enough when preparing for it'. Observe that when a conditional implies a temporal separation between the event or events mentioned in the 'if' clause and the event or events mentioned in the 'then' clause, the same temporal difference is to be retained through all its transformations. Thus, the contrapositive of 'If *a* happens then *b* will happen' is 'If *b* does not happen then *a* has not happened.'

Secondly, an ordinary conditional may, but need not, express a causal relation. Conventional and logical conditionals are non-causal. And it may be said of factual conditionals that they are not always causal. (i) 'If interest rates rise, then savings will increase', seems to express a causal relation in a quite natural sense of 'cause', but (ii) 'If James has red spots, then he has measles' does not. The red spots are not causes of James's measles but form indirect evidence of his having them. And (iii) it may be true that if the rock is from this area, it contains porphyry, even though, from a causal point of view, the location of the rock and its porphyry content both result from a complex prior geologic background. The point is that each of (i), (ii), and (iii) may function as an ordinary conditional because they each claim that the occurrence of the events mentioned by their 'if' clauses would ensure the fulfillment of their 'then' clauses.

One feature that ordinary conditionals never have is that of formal logical validity; if 'If A then B' is an ordinary conditional, 'B' is not deducible from 'A'. If the assertion of 'A' and denial of 'B' is a formal inconsistency, I say that 'A' entails 'B', not that 'If A then B' is a logical truth. I recognize that this involves a substantial departure from common usage, but it is one that I must adopt. Expressions that will qualify as ordinary conditionals in the essay will always have formally contingent confirmations and disconfirmations and they will be deniable without formal contradiction. But I will say that it is only true that 'B' is deducible from 'A' (that 'A' entails 'B') if 'A and not B' is a formal contradiction, and so it would be incoherent on my part to describe entailment statements as ordinary conditionals. It will be argued in what follows, particularly in chapter 6, that the forms I have referred to as 'conceptual' or 'logical' conditionals are not themselves entailment statements, but statements *about* entailments and other logical relations. 'If "A" and if "A entails B" then "B"' is a logical truth, and a conditional, but its denial is not a contradiction and it is not an entailment. If this suggestion seems strange to the reader, I hope that it will come to seem less so as the essay progresses.

A second characteristic an ordinary conditional never exhibits is that of provability. A statement may be said to be provable if in standard discourse it is implied by the statement expressing its confirmation (the statement expressing the observational ground of its assertion). Ordinary conditionals are not provable in this sense. We may observe the fulfillment of the confirmation of a factual conditional, but not the relations in nature that are claimed to exist when we say that the occurrence of one event entitles us to conclude to the occurrence of another event. And we may observe and participate in the acceptance of the use of linguistic expressions but not the relations that establish the conventional or necessary connections between such expressions. Similar observations may accommodate many different conditional explanations. I shall argue that the reason why, for example, what I call 'betting' conditionals and 'disposition' conditionals are not ordinary ones is that they *are* proved by their confirmations. Ordinary conditionals, even if confirmed, are subject to revision in the light of later reconsiderations of the causes that operated in a particular situation or the reasons or motives that underlay a series of human actions. 'Betting' and 'disposition' conditionals, however, make lesser claims about the relations between things; if they are confirmed, they are proved, and nothing that may happen in the future will compel us to reconsider their truth. It is not that these conditionals differ visibly from ordinary conditionals, but rather that they serve different linguistic purposes.

The statement that the conditional I propose to study is not provable will strike some readers simply as false or as at least questionable. And it will be helpful, in illustration of the point I wish to make, to distinguish at the outset this type of conditional from one other type that receives frequent application—the conditional I have just referred to as the 'betting' conditional. When one asserts an ordinary factual conditional of the kind I will be talking about, he is saying that the occurrence of the event mentioned in the 'if' clause, under the circumstances, would ensure that the event mentioned in the 'then' clause will occur or has occurred. This is the conditional that I am saying is not proved by the evidence in its support. Associations between events that give rise to such assurances are not observable, and therefore are not established beyond question by the evidence we rely upon to justify the conditionals that express these associations.

The logic of the 'betting' conditional is quite different. Suppose Smith says, 'If Andy Brown is allowed into the country, then interest rates will rise' . . . I say, 'But that can't be true. Andy's presence in, or absence from, the country has nothing to do with interest rates.' Smith answers, 'I'm not claiming that it does. I don't say that Andy's presence in the country would be sufficient to bring about, or would even be relevant to, a rise in interest rates. I only say that if he is let into the country, rates will rise, nothing more' . . . This is fair enough. If we bet on Smith's claim and Brown enters the country and interest rates rise, Smith wins. I will not be heard to say that he has not established his claim because he has not shown that the rates rose *because* of Brown's entry. What he said would happen on Brown's entry did happen, and that's the end of it.

But if Smith and I were trying to find out how things hang together in the economic world, as a result of physical necessities, social conventions, and human desires and purposes, I suggest we would be saying something else with this conditional. We would be saying, falsely one suspects, that the presence of a certain person in the country would, given the background circumstances, entitle us to conclude that interest rates will rise. We would be making a different statement with the same form of words. And it would not be different because its justification required the direct support of different events. It would be different because, unlike Smith's first conditional, it would be saying something not conclusively established by the occurrence of the events in its support, and our opinion regarding its truth would be subject to review in the light of future discoveries. It is the logic of this second, unprovable conditional that I propose to explore.

I believe that there are several conditionals in common use that differ from the one to be studied here. The disposition conditional ('If you put this in water it will float') and the 'even if' conditional ('Even if you fix the ignition the car will not start') seem to perform linguistic roles dissimilar from those of the type of conditional to be examined. The disposition conditional, once confirmed, seems to be placed beyond doubt. Its confirmation is accepted as its proof; and this is a judgment not to be disturbed by subsequent reappraisals. And the 'even if' conditional does not say that, given the fulfillment of the 'if' clause, the fulfillment of the 'then' clause is assured. It would not be the fixing of the ignition as such that would warrant the conclusion that the car will not start. The 'even if' conditional says, rather, that the fulfillment of the 'if' clause does not warrant the conclusion that the *denial* of the 'then' clause will be fulfilled. I say only that these conditionals have different functions, not that they are always of distinguishable external appearance. The essay does not investigate the logic of these alternative forms, although I will on occasion take the opportunity to mention some of the ways in which the conditions of their truth and confirmation differ from those of what I have called the 'ordinary' conditional (see pp. 53–57).

There is a further type of conditional, to be set apart from the ordinary, the predicates of whose 'if' and 'then' clauses are semantically related in such a way that the 'if' clause actually mentions a situation that is *necessary* to the fulfillment of the 'then' clause or its negation. This type, which I call the 'suppositive' conditional, is illustrated by such expressions as—'If it snows, it will snow heavily' . . . and 'If Smith stands as a candidate, then he will win the election' . . . I mention it here but will reserve further discussion of it until the properties of the ordinary factual conditional are examined in more detail in chap. 1 (see pp. 44–45, 137–8).

The essay also examines the logic of the type of universal conditional that says about all things of a certain category what singular conditionals say about individual things. This is the sort of statement that says that, because of the way events are ordered in the world, if one event or complex of events occurs, one can rest assured that another event or complex of events will occur or has occurred. It is the conditional we variously express by the forms, 'All A is B,' . . . 'If anything is an A, it is a B' and 'Given any arbitrarily chosen thing, if it were to turn out to be an A, it would be a B'. This general conditional is like the singular conditional in that it is not provable; it is about relations between types of event, not essentially about their occurrences, and has reference to all things and not to finite classes of them. It is therefore not illustrated by such limited and provable claims as, 'All the marbles in this bag are red.' Statements of this latter kind are typically asserted of fixed collections of things and can be proved or disproved by examination of the things in question.

On the standard interpretation of the traditional logic of the syllogism and the square of opposition, 'All A is B' is read as '–(∃ x)(Ax and not Bx)' and 'Some A is B' as '(∃ x)(Ax and Bx),' and this interpretation has given rise to the objection, against the traditional logic, that 'All A is B' cannot entail 'Some A is B.' How can the former entail the latter if the former does not, but the latter does, commit us to the existence of things that are A and B? If this objection is justified, then a central principle of the traditional logic, that of subalternation, cannot be sustained. It will be shown (in chapter 3) that, on a certain reading of 'All' and 'Some,' the traditional logic can be taken to be a special application of OC. On this version, the particular form 'Some A is B' does not imply the existence of As and Bs. It is suggested that although most of our uses of 'Some A is B' take for granted the existence of things that are A and things that are B, the particular form does not necessitate this assumption; the truth of the assumption is not a requirement of the truth of the particular statement. That this is so may be seen if we consider 'Some A is B' and 'Some A is not B' in terms of their respective equivalents, 'It is false that all A is not B' and 'It is false that all A is B.' 'It is false that all trespassers on this land will be prosecuted' does not seem to imply that there in fact exists a trespasser on this land who has not been prosecuted. The logic of 'Some A is B' is akin to that of the OC statement 'Possibly if A then B,' which says, 'If a thing is A, there is no assurance on that account that it is not B.' If this view is adopted, it can be shown that all the relations of the square of opposition and all the syllogisms traditionally accepted as valid, are entailments in OC.

The view of conditionality proposed in the essay forces us to consider such related topics as the concept of confirmation, the context of utterance, problems created by conventional relations between predicates, and the nature of entailment. Accordingly, the essay eventually moves into the areas of the logic of confirmation statements (chapter 5), of conventional statements (chapter 4), and of logical statements themselves (chapter 6). The concept of conditionality is shown to be important in these domains as well as in that of factual statements and the distinction I have referred to between conditionality and the notion of formal entailment is developed in chapter 6. The logic that follows, encompasses singular statements and also monadic predicate factual, conventional, and logical statements. I do not try to show in this essay how this logic might be extended to other types of statements—for example, statements containing many placed predicates—nor do I indicate how this could be accomplished. And although the theory provides a decision procedure by which one may determine the entailment relations of all the types of statements considered in the essay, I do not attempt to show how the OC system might be presented as an axiomatic or natural deduction system.

In the text I refer to the views of several writers on the logic of conditionals, but I have not tried to give a systematic criticism or review of the many theories of conditionals that now exist.[1] My object has been to present a coherent theory of a certain type of conditionality based upon a position concerning the role played by direct confirmation and disconfirmation in the deductive relations between statements. I comment on the 'possible worlds' view of conditionals at various places in the text, particularly in relation to the concepts of contraposition and transitivity for conditionals, and briefly, at p. 38, on Ernest Adams's theory that the singular conditional is a probability statement.

Acknowledgments

I am grateful to W. A. McMullen and to the late Gilbert Ryle for their criticism and encouragement. I also thank John Kearns and others associated with the State University of New York Press for helping me to improve on parts of the original manuscript and for the care they have taken in the production of this book.

Chapter 1

INTRODUCTION

The truth functional calculus of propositions is one of the foundation stones of modern logic. It is so eminently reasonable, and its principles perform so important a function in all branches of standard logic, that doing logic without it is unthinkable. And yet, it is a fact that the calculus is only satisfactory as a logical model of a language in which conditionals do not exist. Even such humble assertions as, 'If Smith works hard, then he will get a raise' fall outside the range of statements that can serve as interpretations of the forms of truth functional logic. This would not be an embarrassment if conditionals were marginal characters on the linguistic stage. But our conviction that this is not so, that they are important actors with major roles to play, is evidenced by the many attempts that have been made to find models that properly express their logic. This essay is such an attempt. And its general approach is perhaps best introduced by a consideration of the reasons why the formal relations of conditionals cannot be directly interpreted with the aid of truth functional principles alone. The discussion that follows is concerned essentially with factual conditionals of the type I have called 'ordinary.' The examination of conditionals other than factual will be postponed until chap. 4, which treats of conventional conditionals, and chap. 6, in which I review the characteristics of logical conditionals.

I

The propositional calculus is well designed to serve as the logic of a language addressed to a world of which all one had to say was that certain events have or have not occurred or will or will not occur. And this arrangement would be acceptable if we really thought it the case that the world is a collection of discrete or unrelated events, the occurrence or nonoccurrence of which can be expressed by simple assertions and their denials. Because if reality consists wholly of elementary facts that can, in appropriate circumstances, be determined by observation to obtain or not, and if the sentence letters of the calculus are interpreted by statements or propositions that express these facts, then if the facts are known, all the truths about the world that are capable of rational representation can be embraced in a long conjunction of simple or atomic propositions and their negations of the form 'p.q.–r.–s, . . .' each element of which states the presence or absence of a fact.

(Let us here treat the letters 'p', 'q', 'r' . . . , as sentence letters, to be interpreted by elementary statements. [The logical constants 'not,' 'or,' and 'and' will be represented in the usual way by '–', 'v' and '.' respectively.] Letters placed side by side will be read as conjoined. The capital letters A, B, C . . . will stand for schemata consisting of 'p', 'q', 'r' . . . , and logical constants. Where the verbal expressions 'p', 'q', 'r' . . . are not in question but the situations that make them true, I employ the italicized letters *p, q, r*. . . . Thus, *p* represents the circumstance that would make true the statement that interprets 'p'.)

The formal notion of entailment or logical following in this language presents no difficulty. The proposition B follows from A if and only if it is contradictory to conjoin the assertion of A with the denial of B. And if we want to find out if such a conjunction contradicts itself, we simply inspect it to establish whether it conjoins some element 'p' with its denial '–p'. The point of a theory of deduction is to enable us to keep our discourse at least consistent. Tests of entailment permit us to avoid the manifest contradictions that defeat communication.

If the world is susceptible of atomic description, we will, then, when looking for deductive connections, usually be comparing conjunctive forms with other similar forms to see if one follows from the other; and this account of logical following implies that we will be inspecting for logical contradiction linkages of conjunctions with denials of conjunctions. How is this to be done? A conjunction is false if and only if at least one of its components is false and so the denial of a conjunction is equivalent to a disjunction of the denials of each of the elements in the conjunction. This disjunction is true if one of its disjuncts is true; thus if a conjunctive description is to be incompatible with the disjunction, the marriage of the conjunction to each of the disjuncts of the disjunction must be contradictory. Then to show that the form A.B entails C.D, we must show that

(A.B)(–C v –D)

is a contradiction; and this will be so if each of

(A.B.–C) and (A.B.–D)

is a contradiction. And if contradictions are established, we say that A.B 'truth functionally' entails C.D. By this means we are afforded a definition of logical deduction. A schema G is deducible from H (H entails G) if and only if H.–G is a contradiction. We thus derive entailments from a contradiction by placing part of the contradictory conjunction on the left side of '. . . entails . . .' and the denial of the rest of the conjunction on the right side.

In a universe in which each fact consists in the occurrence of an event and in which each fact can be matched with an elementary statement in such a way that the absence of the fact disproves the statement and the presence of the fact proves it, this theory of entailment satisfies our normal requirements. If, by our tests, A.–B is shown to be inconsistent, then to avoid inconsistency, from the truth of A we must deduce B and from the falsity of B we must deduce the falsity of A.

Although I think, ignoring conditionals for the time being, that this view of entailment is essentially correct as far as it goes, I shall take a moment to propose a modification of it in the light of a prerequisite I hereafter impose on the statements of OC, and to which I shall refer as the 'application requirement'. If a system of logic is to be used as a guide to the deductions permissible in our speech, the expressions standing on the left and right sides of proposed deductions must be capable of interpretation by expressions that refer beyond themselves to situations that may or may not obtain; with respect, for example, to an uncombined statement form, it must be possible to indicate a situation or situations that would confirm it and other situations that would disconfirm them. I have two main reasons for imposing this restriction. First, let us suppose that an expression A does not have any coherent form of confirmation. Then it will not be of use to us to know that A entails B, because, not being able to say what are the circumstances under which the assertion of A would be justified, we cannot say how the entailment could ever be applied, the conditions under which the deduction of B from A would be a sound one. If an expression is to qualify as a candidate for entailment relations we must be able to describe the states of affairs that would confirm it.

But if the left side of an entailment is to be comfirmable, so must the right side, because if A receives support we must be able to say that B receives support to at least the same degree. Further, if an entailment holds between A and B, it must also hold between –B and –A, and so –B and –A must also be capable of receiving support. For this reason, if A and B are to be related by way of entailment, it must be possible to describe the evidence that would support them and also the evidence that would disconfirm them.

Secondly, as we shall see in the course of this chapter, the strategy of OC is to test the entailment relations of expressions by estimating the effect their confirmations and disconfirmations would have on each other; if, then, an expression lacks conditions of confirmation and disconfirmation that are identifiable in principle, we will not only be unable to apply our logic of entailment to this expression, we will have no logic for it at all, for want of capacity to specify the situations that would, were they to occur, act to confirm or disconfirm the expression.

The application requirement is not a feature, like consistency, which is an essential characteristic of any logical system. It does, however, as will become clear as the essay proceeds, occupy an important place in the theory OC. It comes to this. A sentence letter or a sentence or a predicative expression that becomes a statement when supplied with an identifying expression, or a group of such, only has a logic, only has logical consequences, if it can be interpreted by a (simple or complex) statement. And a verbal form will not be a statement unless one can say, or show, what would be its evidential support and what would constitute evidential support for its denial. This is because the logic of a form is given in terms of the relations that form has to the evidence that would support it and the evidence that would undermine it. This is obviously not to say that a form, to be a statement, must *be* supported or infirmed in fact—a conditional, for example, may be true though not in fact supported by wordly events and it may be false though not disconfirmed. The point is just that we must be able to say, of expressions to be interpreted by statements, what the evidence for or against them would be like, were it to occur.

And a similar trouble affects contradictions, which will be identified in the essay as formally inconsistent conjunctions or disjunctions of such conjunctions. They are incapable of application because attempts to confirm one member of an inconsistent conjunction will be found to yield the confirmation of the denial of another member of the conjunction. The member statements of an inconsistent conjunction *could* not be true together. Again, I am not saying that a conjunction, to count as a statement, must in fact be confirmed or disconfirmed; it is simply that we must be able to say of it what events would confirm each sentential member of it without confirming the denial of any other member. And if we dismiss inconsistent or contradictory antecedents, we must also disallow contradictory consequents, because we know that there is no application that would justify the deduction of a contradictory conclusion. And having come this far, we must also exclude tautologous expressions (denials of inconsistencies) from the antecedents and consequents of deductions because, where A entails B, it must also be true that –B entails –A.

But our account of entailment so far permits us to treat contradictions and their denials as members of the left and right sides of deductions. Because 'p.–p.q' is a contradiction we may, by our test, say the 'p.–p' entails '–q' and that 'q' entails '–p v p'. The modified version of entailment I recommend is this. In deriving entailments from contradictions by putting part of the contradiction on the left side of 'entails' and the denial of the remainder on the right, we must take care that the forms on the left and right sides are contingent, that neither they nor their denials are self-contradictory. I am not arguing that contradictions are senseless or without meaning. They are legitimate linguistic expressions to which we can ascribe characteristics. But confirmation of the member statements of contradictions leads to confirmation of the denials of other member statements, and therefore, by the application requirement, we cannot make statements *with* them, although we may say things *about* them. (I realize that this position is contrary to the one most frequently held. Objections to the approach are considered in chap. 6, where this subject is more fully discussed.)

Again, this is not to say that contradictions and tautologies (the latter of which are denials of contradictions) are senseless or without a role in language. They have logical properties that may be characterized by statements *about* them. It is simply that one cannot make a statement with an expression that reduces to an inconsistency or a tautology. The OC method of testing for inconsistency is sketched in sec. 4 of this chapter and stated more precisely in chap. 2.

How, if at all, might conditional statements be fitted to this frame? Can this form of statement we use to report the relations between events—'If he passes his exam, his parents will be pleased', 'If the barometric pressure is rising, the weather will be fair'—can the logic of this form be properly revealed by a conjunction or disjunction of assertions and denials of elementary statements? The answer of classical theory is that we should think of the conditional in terms of what falsifies it. The conditional 'If p then q' is false if it is the case that 'p' is true and 'q' false. If we say that the denial of the conditional is *equivalent* to 'p.–q,' then the assertion of the conditional is equivalent to the denial of 'p.–q', that is, to '–p v q'. It is possible, according to this theory, to bring the conditional within the scope of the propositional calculus—the denial of the conditional becomes a conjunction and its assertion a disjunction. Its truth or falsity may be conclusively determined by reference to the truth or falsity of its elementary components.

Unfortunately, if this version is accepted, '–p' entails 'If p then q', and so does 'q', because, respectively, '–p.p.–q' and 'q.p.–q' are contradictions. Also, given any two states of affairs *p* and *q*, if *p* does not obtain, then one can consistently maintain both 'If p then q' and 'If p then *–q*'. And, whatever may be the case in the world, either 'If p then q' or 'If q then p' is true. But this is to trivialize an important region of speech. The relations between events are just as significant to us as the question of their occurrence. We want to know the circumstances under which things happen—and this is so whether the subject matter is particles of light, cells in the body, commercial responses to movements of prices on the open market, or the behavior and feelings of animals and humans. To ground a claim that 'If p then q' is true on the fact that 'p' is false or 'q' true misses the point of what the conditional is meant to say. A conditional, when applied to the experienced world, attempts to report a dependency between events, a dependency that obtains as a result of physical connections, social conventions, or human purposes; this objective is not fulfilled by the fact alone that our observations show that the assertion has not been disconfirmed.

Our efforts to resolve this problem, and to devise a logic that accommodates the conditional of the type I have been considering, have been hampered, it seems to me, by a legacy of the form of atomism to which I have referred: the belief that singular conditionals, if they are true, are made so by the occurrence of an identifiable event or events. The explanation of what I mean by this statement necessitates a digression. We know that there are countless assertions that we regard as being true or false but that employ concepts that do not take observable instances. There are no observable properties through which such concepts as pressure, energy, force, gravity, and time are immediately instantiated. We cannot reduce our treatment of assertions employing these concepts to statements specifying their observable properties and saying that they are true when these properties are present, otherwise false. If we did we could not make use of such assertions in a language designed to say things about the experienced world—the 'properties' are not there to be observed. We cannot catch glimpses of bits of energy or force or time as we can of rare birds or butterflies.

This difficulty is more radical than that posed by statements about the past or the experience of others. Although we cannot now observe Caesar crossing the Rubicon, we can say what observable happenings would have constituted his crossing the Rubicon and established the truth of 'Caesar crossed the Rubicon'. And although I cannot observe the after-image in another's visual field, I can know what sort of thing he must be experiencing if 'He sees a blue after-image' is to be true. The trouble with bits of time or force is that they cannot be observed by anyone, at any time.

But recognition of this problem has not persuaded us to abandon all discourse that makes significant use of concepts that are not directly satisfiable by instances of observable properties. What we do, as everyone knows, is establish the application of these concepts in terms of statements expressing characteristics that *are* observable. We find ways of measuring, for example, time and force by observing the behavior of things—either natural things or artificial constructions such as dials and meters—which are not themselves times or forces. The extent to which we treat the statements about observables as 'saying the same thing as' the statements about unobservables that they support, will vary with the concepts in question and the purposes to which they are being put. There is no question, however, that if our discourse is to be public, the justification and infirmation of statements about unobservables must be supplied in terms of statements about observables. If we are to have a common understanding of what we are talking about, we must be in a position to refer problems of saying when a thing is so or not so to things and situations having characteristics to which we can all bear witness.

Now the difficulty with factual conditionals of the type I will examine starts with the circumstance that they say things about unobservables. 'If p then q' says that the truth of 'p' would ensure the truth of 'q' and the conditional is true just in case relations between things in the world are such that the conclusion would be borne out by the facts. This, if you will, is its 'truth condition'. But although we may witness the fulfillment or otherwise of the situations to which 'p' and 'q' are addressed, the relations in the world that assure us, if 'If p then q' is true, that 'q' will be true if 'p' is, are beyond the reach of our senses. And this trouble is made acute by the fact that the observation statement that seems to provide the conditional with its most significant direct support cannot be taken to say the 'same thing' as is said by the conditional itself. Consider, for instance, what is usually taken as the strongest evidence for the truth of 'If p then q', the joint fulfillment of *p* and *q*. This by no means furnishes proof of 'If p then q'. Suppose someone predicts that if wages remain constant, investment will increase and that wages do remain constant and investment does increase. It is still open to critics of the hypothesis to argue that the unchanging wage rates were not what induced the increase in investment. They might say, for example, that it was the increased foreign demand for domestic product or declining interest rates that made the difference and that the rise in investment would not have been affected by a rise in wages. The truth of 'If p then q' cannot be extracted, as a matter of logical necessity, from the truths of 'p' and 'q'. Notwithstanding the truth of 'p and q', the truth of the assertion that if *p* occurs, this ensures the occurrence of *q*, may be questioned without inconsistency. The order of things is for us of final concern; it is appropriate that the logic of the language we use to describe this order does not force us to draw dogmatic conclusions from finite cases.

Although the problems posed by unobservables for atomistic theories are well recognized in the philosophy of science, we have been slow to concede (perhaps influenced by the simple perfection of the propositional calculus) that the puzzles of the logic of conditionals may have their origin in a similar source: the fact that the relations that conditionals assert are not susceptible of direct observation. We have wanted to regard proper singular conditionals as reducible to atomic statements expressing the occurrences of events and truth functional compounds of such statements. This is a view of conditionals that is breaking down; but it is one that still influences our thought on the logic of conditionals. Even some who reject the truth functional model of the singular conditional propose systems in which the truth of the conditional can be equated with the joint occurrence, in certain possible worlds, of the event mentioned in the 'if' clause and that mentioned in the 'then' clause. (I discuss an example of this practice in sec. 5.)

I propose to show how we can develop a logic of ordinary conditionals that recognizes that conditionals cannot be reduced to statements, having a truth functional logic, about the occurrences and nonoccurrences of events, but that also does justice to the role played by such occurrences and nonoccurrences in the assessment of the truth or falsity of conditionals and their logical relations.

Let us try to see why the reduction of conditionals to statements about observables cannot take place, in the hope that we can find a way to bridge the gap between these two kinds of statements. I shall assume that formal logic is about statements and that it concerns the formal effect of the truth or falsity of one statement on the truth or falsity of another. A form of words is not a statement by virtue of being a grammatical sentence. A sentence becomes a statement when it identifies some thing or things and makes an assertion about it or them. The sentence 'Smith is in London' becomes a statement only when it identifies a particular Smith. Statements are true if what they say obtains, otherwise false, and are equivalent if they say the same thing. Formal logic interests itself in the relations statements have to each other in virtue, not of what they say, but of the fact that they say *something*. For the purposes of formal logic, we know of a statement not what it says of the things to which it refers, only that whatever it says is true or false.

Let us suppose for the moment that the impossible were possible and that we had gathered up all the past, present, and future occurrences of all events in all the universe (including among 'events', states, states of affairs, and situations) and that we then correlated with these occurrences all the singular elementary statements and their denials. We could then tell which elementary statements were true and which false and we would end up with a long conjunction of elementary statements we knew to be true. A formal logic of the type I have described would permit us to express all the logical relations it would be possible for this set of statements to have. Now among these fulfillments, these occurrences of events, we would not find the connections *between* events, the arrangements in nature or society that would make it true that if one event has occurred, another event will have occurred. This is because the connections between events are not themselves events that can have occurrences. So if singular statements are only about events the occurrences of which can, in principle at least, be directly detected, and compounds of these occurrences, we will have nothing to which conditionals can be attached. There will be no 'something' that conditionals say; they will not be about anything, will not be statements and will have no logic. But we do constantly make conditional assertions—in law, science, morals, and literature, not to mention ordinary speech—and so we must either allow that this part of our common discourse does not accomplish what it purports to accomplish: that 'If this pen is in my hand, then the earth travels around the sun' is true because it says no more than 'Either this pen is not in my hand or the earth travels around the sun'; or we must show how the conditional is related to the world: how we can justify the claim that a conditional is true or false notwithstanding that the relation between things, which must obtain if the conditional is to be true, is not the occurrence of an event. If we are to treat 'If p then q' as capable of being true or false even though the situation that would make it true is inaccessible to sense, we must be able to state the situations that are so accessible, and which, although they do not prove the conditional, do provide justification for its assertion.

I shall use the words 'confirmation', 'fulfillment', and 'support' to express the notion of direct evidence for the truth of a statement, whether that evidence proves the statement or merely strengthens the grounds for its acceptance. 'Disconfirmation' and 'infirmation' will stand for the direct evidence for the falsity of the statement, again whether that evidence disproves the statement or simply makes its truth less likely. Situations which, if they obtain, are conclusive, which put questions of the truth or falsity of the statement beyond dispute, will be said to 'prove' or 'disprove' the statement. In the text as a whole I shall often refer not only to the evidence, but also to the phrases that describe the evidence, with the words 'confirmation' and 'disconfirmation' (although it would be more correct to refer to the latter with the words 'confirmation statement' and 'disconfirmation statement'), in the belief that no ambiguity, in context, will result from this more relaxed usage.

The difference in the way in which certain factual statements are related to the statements that support or infirm them may be brought out in this way. Let us say that a factual statement asserts that a certain 'objective state' obtains. Let us say also that the evidence for the presence or absence of the objective state may be given in terms of statements, to be called 'observable aspect' statements, which are singular statements expressing the occurrences or otherwise of the events the statement expressing the objective state is about. The objective state is what the original statement asserts of the events it is about. The observable aspect of the objective state is described by direct evidence statements that express the occurrence or nonoccurrence of the events in question. Direct evidence statements may report events that are directly observable, as in 'The cat is on the mat', 'This is blue', or they may be taken, within a particular discourse, as equivalent to sets of observation statements—thus we tell time by reference to clocks and the temperature by reference to thermometers. An observable aspect statement is one for which there is a criterial set of observations from the fulfillment of which the truth of the observable aspect statement may be inferred. The group of observations forming this 'criterial set' is not immutable. We may, for different purposes, adopt different tests for time and temperature. It is sufficient, if a statement is to act as an observable aspect statement, that there be a criterial set of observations that determines the truth of the statement and also a criterial set that determines its falsehood, at least within the context of a particular discourse.

The expression 'observable aspects' here has to do with events whose occurrences are recordable in principle, not necessarily in fact. Thus 'Caesar crossed the Rubicon' has associated with it a direct evidence statement that describes the occurrence of the event that is the observable aspect of Caesar crossing the Rubicon, although this event cannot now be observed. Nor is it even necessary, if a statement is to be a proper observable aspect or evidence statement, that it has been or will be fulfilled—that the appropriate observations have been or will at some time actually be made. It is requisite only that the observations could be or could have been made by a person in the right position at the right time. I do not attempt to deal in this essay with the more complex situations that are introduced when the statements that report occurrences of events (for instance, statements about the activities of subatomic particles) are not treated as criterial and are themselves insufficient to *prove* those occurrences.

A factual statement asserts that an objective state obtains. If the objective state is expressible by a direct evidence statement of the type I have described, then the state and the observable aspect are the same and the statement itself, the 'parent' statement, is the same as the statement that reports its fulfillment, which reports, that is to say, the occurrence of the event that certifies its truth. In such case the parent statement is said to be 'elementary'. The fulfillment of the parent statement, 'The cat is on the mat' is expressed by 'The cat is on the mat', the truth of which proves the parent. There are statements, however, whose objective states are not observables or groups of observables and that cannot be said to be equivalent to their observable aspect statements. "Something is an auk', and 'Something is bigger than everything other than itself' are statements of this type. There is no finite set of observation statements that would imply the falsity of (disprove) the first and no such set that would imply either the truth or falsity of the second. But these statements are related to their observable aspect statements in definite ways. The first is proved if we find an auk, although it may be true even if no auk has been found; and although it is not disprovable, nevertheless if we can find no auks, we are said to have direct evidence that there are none. If we observe something bigger than anything else we can find we have evidence (though not conclusive evidence) for the truth of the second; yet if we find something exactly the same size as this large thing, although we have some evidence that there is nothing that is larger than everything other than itself, we must still allow the possibility that there is something larger still that *is* larger than everything other than itself. Such statements, then, although not elementary and not equivalent to the observable aspect statements of their assertions or denials, nevertheless satisfy the application requirement in that it is possible to describe situations that would confirm them and also situations that would disconfirm them; and they also have a logic that may be elucidated in terms of the occurrences or otherwise of the events they describe.

I will argue that the logic of conditionals may be presented in this way. Conditionals are not provable; they are not implied by the observable aspect statements that express the occurrence or nonoccurrence of the events they are about. But they satisfy the application requirement because it is possible to describe situations that would confirm them and situations that would disconfirm them in terms of the occurrences or otherwise of the events to which they have reference. The objective state asserted by the conditional is a relation between events that ensures that, when the 'if' clause has been fulfilled, the 'then' clause has been or will be fulfilled. This relation between events might be referred to simply as the 'truth condition' of the conditional were it not for the fact that in most contexts 'truth condition' has come to imply a situation that may be observed to hold or not to hold. But what I have called the objective state, the situation that makes a statement true, may not be, and in the case of the conditional is not, a state capable of being observed, and for this reason I have given it a different name. Confirmations and disconfirmations of a conditional, we shall see, are conjunctions, or disjunctions of conjunctions, of direct evidence or observable aspect statements having a truth functional logic. These direct evidence statements do not constitute a reduction of the conditional that asserts the objective state, but it will be through an understanding of the relations of these to conditionals that we will learn to construct the logic of the latter. And the determination of the logical form of the confirmations and disconfirmations of conditionals will be vital to the formulation of their logical relations, because the entailment relations of conditionals will be established on the basis of the form of their confirmations and disconfirmations and the relations these bear to the conditionals whose observable aspect statements they are.

II

In the literature on conditionals, the 'if' and 'then' clauses are often given the names 'antecedent' and 'consequent'. I will, in general, reserve this latter pair of expressions for the left and right sides of entailment statements, for situations, in other words, in which the consequent is formally deducible from the antecedent. But, as I have noted in the Preface, the 'then' clause of an ordinary conditional is never formally deducible from the 'if' clause, and I have therefore adopted a different terminology for conditionals. Consider a conditional of the simplest form, 'If p then q'. I will refer to the clause that follows the 'if' as the 'condition' or 'condition clause' and the event or circumstance expressed by that clause as the 'conditioning' event or circumstance; the clause following the 'then' will be the 'conclusion' or 'conclusion clause' and the event or circumstance expressed by that clause the 'projected' event or circumstance. Although, as I have said, I generally reserve 'antecedent' and 'consequent' for the left and right sides of entailments, I will sometimes use these words in place of 'condition' and 'conclusion' when discussing the views of authors who use 'antecedent' and 'consequent' to describe the parts of conditionals.

How do we gather evidence for or against a singular conditional of this form? Suppose I see Henry on the green hunched anxiously over his putter and about to attempt a twelve-foot putt. I say, 'If I startle Henry, he will swear'. The first thing to notice is that the conditional would be disproved by the occurrence of the conditioning event and nonoccurrence of the projected event. If I startle

Henry and he fails to swear, my conditional is laid to rest. Thus the conditional has this link with experience: the unobservable relation that is said to exist by 'If p then q' is conclusively shown to be absent by the truth of the condition and falsity of the conclusion. We can thus form the disconfirmation, and disproof, of a conditional by conjoining the assertion of the condition with the denial of the conclusion.

The second thing to notice is that the truth of the conditional is not deducible from the falsity of its disconfirmation. If I do not startle Henry, or he swears, my conditional cannot be disconfirmed; but I am not entitled to conclude from this that the conditional is true, that if I were to startle Henry, he would swear.

And this leads to a third observation—the conditional may be false even though it has not been disconfirmed. It may be false that if I startle Henry he will swear even though it is not the case that I have startled Henry and he has not sworn. To change the illustration, it is false that if I do a rain dance, it will rain, and this is so notwithstanding that I have not performed a rain dance that has been unaccompanied by subsequent rain.

The conditional is then a statement whose objective state is an unobservable relation between its conditioning event and its projected event; and the objective state of its denial is the absence of such a relation. The observable aspect of the statement's denial consists in the occurrence of the conditioning event and the nonoccurrence of the projected event, and these occurrences disprove the conditional. But though a conditional be false, the observable aspect of its falsity need not have been fulfilled; we have seen that a conditional may be false though not disconfirmed.

What of the observable aspect of the truth of the conditional 'If p then q'? We can at least say that its form is not the logical equivalent of the denial of its disconfirmation. The denial of 'p.–q' is 'p.q v –p.q v –p.–q'. Of these three possible situations, '–p.q' does not seem to afford any justification for the assertion of the conditional. If I do not startle Henry, but he swears in any case, I am left with no evidence that my startling him would have induced his swearing. The commonsense approach of those who do not have an interest in the defence of the truth functional theory of conditionality is usually to allow only 'p.q' as the confirmation of 'If p then q'. I startle Henry and he swears: I may conclude that my conditional has received the only form of direct confirmation it is capable of taking. But I suggest that in saying this I have ignored the role of the background against which these confirming events have occurred, the situation in which I have not startled Henry and he has not sworn, and that it is really the change from the absence of the startling action and Henry's verbal piety to the startling and the swearing that helps to justify the conditional assertion. Suppose Henry swore constantly while on the green (to relieve tension, let us say): then we would not say that my startling him while he swore bolstered the claim that he swore because I startled him. It is a change in the situation, accompanied by another change, which confirms the singular conditional. This movement need not be from the nonoccurrence of events to their occurrence. For example, 'If you stop shaking your stick at the dog, it will stop barking' is confirmed by a movement from stick shaking and barking to nonstick shaking and nonbarking. It is not the direction of the movement from one state to another but the fact that the states occur in succession, which results in significant confirmation of the singular conditional.

Let us notice that this form of confirmation is not affected if the condition clause refers to an event that occurs after the event to which the conclusion refers. 'If John is cheering up, then Lucy must have called' is confirmed by a period during which John is morose and Lucy has not called and another period during which Lucy has called and John cheers up.

I will say, where '(i)' and '(ii)' represent different temporal parts of the situation to which 'If p then q' applies, that each of '(i)–p.–q' and '(ii)p.q' is a 'form of confirmation' or 'confirmation part' of 'If p then q'. But each by itself provides only nominal confirmation for 'If p then q'. A nut and a bolt form two parts of a fastener. Each is essential to the fastening action, although neither is itself a fastener. The nut is not a weak fastener, nor is the bolt. Similarly, the situation expressed by '(i)–p.–q' does not supply half the justification, '(ii)p.q' the other half, for the assertion of the conditional. It is the combined truth of '(i)–p.–q' and '(ii)p.q' that constitutes the direct justification for the singular conditional 'If p then q'. But just as a nut and a bolt are each part of the fastening device, so each of '(i)–p.–q' and '(ii)p.q' is a part of the confirmation of 'If p then q'. The phrase, "full confirmation of 'If p then q' " means the confirmation of the conditional during the phases of the occasion in question by the different confirmation parts '(i)–p.–q' and '(ii)p.q'.

But how can one say that the nonoccurrence of two events and also the occurrence of those events can serve as support for the same assertion? The forms 'p.q' and '–p.–q' are contraries and cannot both be true. I seem to be saying that the conditional must be confirmed by an inconsistency. And yet there was nothing inconsistent in the argument that showed that the confirmation of 'If I startle Henry, he will swear' consisted of one state in which Henry, unstartled, did not swear and of another in which Henry swore on being startled. To avoid inconsistency we must take care to say that the different forms of confirmation occur during different phases or segments of the single occasion of which the conditional is said to be true. It is not inconsistent to say, in reference to one phase or stage of the occasion with respect to which 'If I startle Henry he will swear' applies that I do not startle Henry and he does not swear, '(i)–p.–q', and in reference to another phase that I do startle him and he swears, '(ii)p.q'. The conditional asserts a relation for which its confirmation parts supply the evidence; but what the conditional asserts is not reducible to one or both of its confirmation parts. In demonstrating the logical relations of a conditional with other statements, I will show the logical effect of each confirmation part in turn on those other statements.

A singular conditional is not a covert universal about all times or phases. 'If p then q' does not imply that always, if p then q—that, for instance, whenever Henry is startled, he swears. It limits itself to the claim that, on the occasion to which it is addressed, the events it mentions are related in a certain way. And this is a claim that would be supported by, but is not reducible to, occurrences, during phases of the occasion, of the events appropriate to its confirmation.

It may happen (usually does happen) that the observable aspect of a conditional is not partially or completely fulfilled by events. And yet a conditional may be true, even though not confirmed, or not confirmed in each alternative way. Thus 'If p then q' may be true notwithstanding that 'p.q' or '–p.–q' or '–p.q' is true throughout the occasion to which 'If p then q' applies. In such case, however, 'If p then q', even if truly expressing a relation between *p* and *q* that would make it correct to conclude that 'q' is true if 'p' is true, is not supported by the sort of evidence I have called 'direct'—occurrences of the events to which the conditional has reference. It is clear that events that are not logically related to the events referred to by a conditional may influence the formation of our beliefs about its truth or falsity. I may believe that if I ring the Smiths' door bell, someone will come to the door, because I see that a light is on in the living room. It is not this indirect form of evidence, however, which I will use as a guide to the logic of statements, but rather the evidence delivered by occurrences of the events of which the statements speak.

We should notice that even the confirmation of a conditional in each alternative way does not prove the conditional. Henry swore when startled; but he may not have sworn because he was startled. It is possible that he swore because he was just then stung by a bee or he remembered that he had left his car lights on or had neglected an appointment or for an unlimited number of other reasons.

This analysis indicates that an appropriate logic of conditionals will be such that 'If p then q' will be inconsistent with its own disconfirmation but it will not entail the truth of either of its confirming situations. Also, although disconfirmation disproves the conditional, confirmation does not prove it—we cannot deduce the truth of the conditional from the fact of its confirmation in each alternative way. And although the falsity of the conditional is proved by its disconfirmation, the conditional may be false even if not disconfirmed. The confirmation of a statement is the disconfirmation of its denial and its disconfirmation the confirmation of its denial. Thus the confirmation of 'If p then q' disconfirms 'It is not the case that if p then q' and the latter is confirmed by what disconfirms 'If p then q'. The form of conditional to be presented in chap. 2 will exhibit all these logical characteristics.

To prevent misunderstanding, I should add that although I believe that the 'phased' or 'segmented' character of the full confirmation of a conditional is an important key to its logic, I do not suggest that we use conditionals on the basis of a conscious application of this feature of their confirmation. Most people have not addressed their minds to the possibility that a conditional may have more than one form of confirmation; and many, when confronted with the idea, may reject it. What I propose is that our use of language, as distinguished from our awareness of the structure of that use, does recognize the role of phased confirmation and that attention to this kind of direct support helps us to explain the logical design of conditionals.

I have been saying that conditionals assert the presence in the world of certain causal and other arrangements among things and that the events that support them, if these events occur, do not prove that the arrangements actually hold. But if this is true of the ordinary conditional, then in asserting the conditional we are forced to assume that these arrangements are in some sense 'there' in the world—are true of it. Because if we say that an expression can be true or false and that what it says is that X obtains, then the expression is only true if X does obtain. And this is so even if we concede that the evidence for the presence of X is never conclusive. We allow that the world may not be patterned in the way we suppose it to be, even that in some instances and in respect of some areas of study, there may not *be* patterns that would support true conditionals; but we believe that the world does embody certain patterns that it is worthwhile trying to represent in conditional statements, always conceding that the evidence we may collect in support of these conditionals cannot certify their truth. There is as well a further assumption that seems to underlie our use of conditionals of the type I have described. It is the conviction that, except for borderline cases, the direct evidence for the truth of observable aspect statements is conclusive; we are always more certain of the truth of observation statements than we are about statements that assert connections between events. It is thus that we are able to disconfirm conditionals. If we can have no more confidence in the truth of 'This glass is released' and 'This glass does not shatter on the floor' than we have in 'If this glass is released, it will shatter on the floor', we cannot say that the conditional is disconfirmed if the glass is released and fails to shatter.

There is another feature of a language that I should mention and this is that in trying to come to terms with its logic we assume that language is a rational, intelligent, not a causal system. In logic we wish to know the criteria of correct uses of expressions and these criteria are independent of the causes of utterance. Circumstances may cause the occurrence of an utterance that is incorrect and the circumstances that would make an utterance correct may not cause it to occur. The question for the logician is the normative one; 'What are the rational conditions of the use of expressions?' And if he asks this question he supposes that there are answers, that there are uses of language that are correct as distinguished from those that simply occur. Knowing the cause of a usage that happens to be correct is not the same as knowing that the usage *is* correct. The causes of verbal behavior are not relevant to logical inquiry.

It should be noticed also that the conditional statements of a language are not always addressed to associations ultimately governed by causal relations. Sometimes in saying that if one expression is applied, another expression is to be applied, we assert that these expressions are related not by (physical or psychological) causes, but by conventions grounded in a normative system that is fundamental to the language's successful operation (see the discussion of conventions in chap. 4). Questions of the logical relations of expressions may in addition be directed to a further noncausal issue, the problem of what, given the acceptance of an assertion, it is then rationally possible to say; the question of where the line between consistency and contradiction is to be drawn. (The logic of logical statements is examined in chap. 6.) Thus it is not only that our language embraces, or makes use of, normative, noncausal systems. Such systems often themselves form the subject matter of logical inquiry.

To the extent that OC accurately reflects the logic of ordinary language conditionals, it would seem that in using these conditionals in ordinary speech we commit ourselves to acceptance of certain philosophical positions that we may or may not regard as valid when we consider these positions from points of view other than that of the logic of natural language. I will not attempt in this essay either a description or a criticism of the ontological commitments that may be forced upon us by our use of ordinary conditionals. My object is to find out what the logic of ordinary conditionals can be said to be, and I do not think that a depiction of the ontological underpinnings of this logic is necessary to the success of the undertaking. In fact, a concern with ontology may make perception of logical designs more difficult by inclining us not to see patterns having consequences inconsistent with favored ontologies. I believe that the great testing ground for the logic of natural language consists in our rejections and acceptances of formal relations between statements in ordinary speech and that if our observations of the connections we impose on the statements of regular communication are to assist us, we must behold the practices that typify our own speech with clear eyes.

III

The occasion to which a singular conditional applies can be thought of as the period within which it may be supported by all the direct confirmation it is capable of receiving. Such periods may extend over long stretches of time. The period during which 'If you spray the apple trees, the apple crop will increase' receives confirmation in each alternative way will include at least two crop seasons, during one of which the trees have not been sprayed and the crop does not increase, and during the second of which the trees are sprayed and we have the opportunity to observe what happens to the crop.

A singular conditional may be true, I have argued, though not confirmed; and it may be false even though confirmed in each alternative way. Also, conceptual considerations may make it impossible, in principle, for a singular conditional to be confirmed in each alternative way. 'If Gordon Liddy had been born a Russian, he would now be a Communist'[1] is an example of a statement that cannot receive full direct support. In the circumstances that obtain, one of its confirmation parts is fulfilled. Gordon Liddy is not a Communist and was not born a Russian. In other circumstances the other confirmation part might have been fulfilled: if Liddy had been born a Russian and had become a Communist. But, whatever the circumstances, the statement cannot be confirmed in *each* way. The confirmation in each way of 'If p then q' requires that each of *p* and *q* is an event that can (logically) occur during one part of the occasion or period to which the conditional refers and not occur during the other part. But a person having been born a Russian implies that he was never born a non-Russian. This is very different from the case of Henry being startled or not startled. My not startling Henry at one moment does not prevent me from startling him the next. It is clear that indefinitely many singular conditionals have the disability of the singular conditional about Liddy. 'If the Yankees get the first run, they will win the game' will be confirmed in one way if the Yankees get the first run and win and in another way if they lose having failed to get the first run. But it is not possible for them to get the first run and not get it in a single game, or for them to win a game and not win it. And the problem may arise as a result of technological rather than conceptual considerations. 'If this is an emerald, then it's green' can be confirmed in one way; but I am unable to change the thing's color to see if it stops being an emerald. Thus even though the confirmation of a legitimate conditional in each alternative way is a formal possibility, it may not, because of the circumstances or the meaning of the expressions it contains, always be a logical possibility, in a broader sense of the word 'logic'. In such cases the singular conditional cannot be confirmed in a way that provides substantive support for the conclusion that the conditional is true. (See the discussion in chap. 4 of the influence of conventional meanings and of contextual assumptions on the logic of factual statements.)

Although conditionals like the one about Liddy are limited in the extent to which they can receive direct confirmation, they may be powerfully supported by statements from which their truth may be inferred. Some singular conditionals, for example, are deducible from universal ones. I show in chap. 3 that certain singular conditionals are deducible from universal conditionals of the form 'All things which are h are g'. And because we may gather evidence for universals from different occasions as well as from different segments of the same occasion, the opportunities for obtaining support or infirmation of a universal conditional are much greater than is the case for the singular conditional. If we were to find, for instance, that 'All those born Russian become Communists' is confirmed by a large number of observations of Russian-born Communists and of non-Communists who are not Russian-born and we find no cases of Russian-born non-Communists, we would accumulate strong support for the universal and also, by entailment, for 'If Gordon Liddy had been born a Russian, he would today be a Communist'.

It will be important to keep in mind that many actual conditionals have disabilities of the types I have mentioned; but the formal relations of conditional statements will be traced on the assumption that each conditional in a formulation is capable of receiving confirmation in all the alternative ways its structure may imply.

As will be seen in chap. 3, the singular conditional does not entail any corresponding universal. Confirmation of the singular conditional confirms the universal from which it is deduced, but disproof of the universal does not necessarily disprove the singular. The universal purports to be true of all things and is disconfirmed if shown to be false of anything. The singular conditional is only disconfirmed if shown to be false of an identified thing or things on the occasion to which the conditional is to be applied. If, however, we believe that factual conditionals, whether singular or general, are true, if true, because of a causal relation between conditioning event and projected event, the falsity of the universal will incline us to doubt its singular form. We will offer the disconfirmation of the universal as a reason for denying the singular conditional, even though the latter may have been confirmed in each alternative way. And yet some will hold that a conditional can be true notwithstanding that any universalization of it may be false. This will be so if, in relation to human behavior, for instance, one believes that the connection that justifies the conclusion that an action will occur if certain conditions are fulfilled is based on intelligent purposes rather than on mechanical causation. On this assumption one would wish to say, for example, that 'Given the present circumstances, if Smith is threatened, he will not pursue his inquiry' may be true although, 'Under similar circumstances, if any person is threatened, he will not pursue his inquiry' may be false. One will say that this may be the case when the condition affords a 'reason' for, rather than a cause of, the occurrence of the ensuing action. 'The statement that one person did something because, for example, another person threatened him, carries no implication or covert assertion that if the circumstances were repeated the same action would follow.'[2] On this view, 'If Smith is threatened, he will not pursue his inquiry' does not even imply the partially general statement, 'Whenever Smith is threatened, he does not pursue his inquiry.' However this may be, it is not, in OC, a deductive step from the truth of a singular factual conditional to the truth of one of its universalizations. The question of the nature of the relation between the conditioning event and the projected event asserted by the factual conditional is a philosophical not a logical issue.

We should perhaps remind ourselves at this stage that while I have couched the discussion so far in terms of factual conditionals, it must not be thought that the relations that underlie the sort of conditionality of which I treat in the essay are always factual. As I indicated in the preface, not every conditional of the type to be studied is dependent for its truth on the existence of relations between conditioning event and projected event that ensure the fulfillment of the projected event on the fulfillment of the condition as a matter of fact. The relations between events may also be either normative or rational. We shall see in chap. 4, for example, that acceptance of the applicability to a situation of one predicate may give rise to a commitment, through the operation of a social rule, to accept the application of another predicate. And in chap. 6, we find that the relation between conditioning circumstance and projected circumstance may be such that, on acceptance of the conditioning circumstance, being some logical characteristic of an expression, rejection of the projected circumstance would make our discourse unintelligible. The thread that binds these different types of conditional is the fact that if they are true, they are so because the fulfillment of the condition is a sufficient ground for a particular conclusion, although, because the relations that obtain between the subjects of condition and conclusion cannot be ascertained with certainty to exist, the evidence in support of the conditionals must fail to guarantee their truth. The claim that the relations *do* obtain is the claim, in the case of the factual conditional, that fulfillment of the condition justifies us in concluding, as a matter of fact, that something will occur or has occurred; in the case of the conventional conditional, it is that fulfillment of the condition justifies us in concluding that a usage ought to be, is required to be accepted;

and where the connection between condition and conclusion is rational or logical, that fulfillment of the condition makes the conclusion inescable, deniable only at the risk of the dissolution of communication. The conditionals that occupy the stage in chap. 2 and 3, however, are without exception factual.

IV

The logic of factual conditionals that I am approaching is one in which the logical relations between statements are determined by the relations direct evidence statements bear to the statements they support and to each other. Some statements are capable of proof but not of disproof and others are disprovable but not provable. To summarize, elementary statements are provable and disprovable; their confirmations and disconfirmations are equivalent to the statements themselves and their denials. The confirmation of a statement is the disconfirmation of its denial and its disconfirmation the confirmation of its denial. In the case of nonconditionals, their confirmations are equivalent to the denials of their disconfirmations. Conditionals have alternative forms of confirmation and are disprovable but not provable. As we have seen, the confirmation of a conditional is not equivalent to the denial of its disconfirmation. Denials of conditionals take alternative forms of disconfirmation and are provable but not disprovable. There are other possibilities; for example, 'For every x, there is some y such that xRy' is neither provable nor disprovable. But I shall not be concerned with such additional varieties in this essay.

The account of the effect on conditionals of the direct evidence for and against them will be incomplete, however, unless we can arrive at a principle upon the basis of which the forms of confirmation and disconfirmation can be constructed. We have seen that the confirmations of 'If p then q' are '(i)p.q' and '(ii)–p.–q'. But how does one construct the confirmations of more complex conditionals such as, for example, 'If p and q then r' or 'If p or q then r or s'? What seems to me to be the answer is that conditionals are confirmed by arrangements that nullify a fulfilled risk of disconfirmation. And what I mean by this is briefly as follows.

An expression that succeeds in making a statement is vulnerable to confirmation and to disconfirmation. A statement, if elementary, is confirmed if, on the occasion to which it has reference, a certain event occurs or has occurred; otherwise it is disconfirmed. The risk that an elementary statement will be disconfirmed is always unconditionally present; the occurrence of an event that either proves or disproves the statement is always inescapably forthcoming. In the case of the conditional, however, the risk of its disconfirmation may not be fulfilled, in which case the events that occur on the occasion in question will neither confirm nor disconfirm it. If, for instance, '–p.q' is the case throughout the occasion to which 'If p then q' applies, the conditional will not be confirmed in either alternative way, nor will it be disconfirmed. Why do we say that '–p.q' does not confirm the conditional? I suggest that it is because '–p.q' cannot be read as a response to the risk of the falsification of 'If p then q'. 'If p then q' would be disconfirmed by 'p.–q' and '–p.q' does not contain either element of the conjunction that would disconfirm the conditional. A form that confirms a conditional is one that can be seen to avoid the disconfirmation of the conditional it supports in that it contains all but one of the elements that make up the disconfirmation of the conditional and the negation of the remaining element; '–p.q', which contains neither 'p' nor '–q', is not a confirmation of 'If p then q'; but that conditional is confirmed by each of '(i)p.q' (which contains 'p') and '(ii)–p.–q' (which contains '–q'), where (i) and (ii) stand, as before, for different phases of the occasion. In accordance with this principle, 'If p and q then r', with disconfirmation form 'p.q.–r', is placed in a state of risk by the forms 'p.q', 'q.–r', and 'p.–r'. And if each state of risk is fulfilled during a different segment of the occasion in question, then 'If p and q then r' will be confirmed in each alternative way by '(i)p.q.r', '(ii)–p.q.–r', and '(iii)p.–q.–r'. This approach will enable us to construct the confirmations and disconfirmations of singular conditionals and also of monadic general ones of whatever complexity.

But to say this about the confirmations and disconfirmations of conditionals is of no immediate help in charting the entailment relations conditionals have with other conditionals and with nonconditionals. A entails B if A and B are contingent and if A.–B is inconsistent. And A.–B is inconsistent, I have said, if the truth of A and falsity of B issue in a contradiction. But suppose we wish to show that it is the case that

'If p then q' entails 'If –q then –p'?

'If p then q and –(If –q then –p)' is not, on the face of it, inconsistent. If we broke these expressions down, in the truth functional way, by putting '–p v q' for the first and '–(q v –p)' for the second, we would have '(–p v q) (–q.p)', which would yield the disjunction

'(–p.–q.p) v (q.–q.p)',

each conjunctive member of which is inconsistent. But the truth functional counterparts do not retain the conditional properties of the originals, and so this device does not furnish an adequate guide to their entailment relations. We cannot express, in terms of truth functional formulations that can be known to be true or false, the truth conditions of conditionals. These truth conditions are relations between things that cannot be observed to obtain or not. But the direct confirmations and disconfirmations of conditionals *can* be represented by truth functional compounds of elementary statements. Let us see, then, what result we get if we seek to establish the inconsistency of A.–B in terms of truth functional relations among the confirmations and disconfirmations of A.–B. And let us examine in this light

'If p then q and –(If –q then –p)'

for possible inconsistency. I have argued that a conditional receives significant direct support only if confirmed in each alternative way—if states of affairs exhibit appropriate responses to each risk of disconfirmation to which the conditional may be subject. Accordingly, we must divide the occasion to which the conditional applies into as many different segments or phases (to be indicated by (i), (ii), . . .) as the conditional has potential states of risk. Let us call the letters 'p' and 'q' and their denials the 'elements' of the 'components' 'If p then q' and '–(If –q then –p).' The disconfirmation of 'If p then q' is 'p.–q' and its states of risk are therefore represented by the forms 'p' and '–q', each element signifying the fulfillment of all but one of the elements contained in 'p.–q'. If 'p' is fulfilled during phase (i), then to avoid disconfirmation the situation must support the truth of '(i)p.q' and if '–q' is fulfilled during another phase (ii), the situation must support the truth of '(ii)–p.–q'. Now observe that if during the occasion in question the situation allows the confirmation of 'If p then q' in the way described, then 'If –q then –p', having the same disconfirmation and therefore, as a conditional, the same confirmations as 'If p then q', will also be confirmed in each alternative way. Thus the confirmation of 'If p then q', when fully realized, is not only inconsistent with the confirmation of '–(If –q then –p)', it is inconsistent with the failure of 'If –q then –p' to be confirmed in each alternative way. Also notice that if '–(If –q then –p)' should by confirmed, and proved, by the realization of '–q.p' during the occasion, then the denial of 'If p then q' will also be confirmed, and proved. So we can say of

'If p then q and –(If –q then –p)'

that the confirmation of 'If p then q' would confirm the denial of '–(If –q then –p)' to its fullest extent and the confirmation of '–(If –q then –p)' would confirm the denial of 'If p then q' to the same extent. Also, the proof of the provable component disproves the other. This is essentially the account I wish to give of inconsistency as it applies to conjunctions having only two components. A.–B is inconsistent if the confirmation of A confirms B to its fullest extent and the confirmation of –B confirms –A to its fullest extent, provided that if one component is provable, its proof must disprove (prove the denial of) the other. The more general definition offered in chap. 2 modifies this account in order to permit its application to many-membered conjunctions.

I do not maintain that this view that inconsistency in a formulaltion should be mapped in terms of the direct confirmations and disconfirmations of its components is the only plausible one to adopt where the object language of a logic includes statements, such as ordinary conditionals, which make claims that go beyond what is said by the statements that supply their direct support. Nor do I claim that the only way to reach a proper logic of conditionals is through the utilization of the distinction between what statements assert, their objective states, and the situations that support them, their observable aspects. Differently based systems may appear that will do the job as well or better. And yet, because it is only if A holds that we can apply the rule that A entails B to infer that B, and because A may be exemplified by statements, in particular conditionals, the public or observable support for which must tell us less than what is asserted by the statements themselves, I suggest that there is something to be said for a method that enables us to apply the entailment of B by A on the basis not of the fulfillment, with respect to A, of some probative state to which we are unable to gain access, but on the basis of what, in principle, might be observed about A.

The situation is this. We may not have access, even in theory, to the objective states that would make A and B true. And where this is so, it is unhelpful to say, without more, that A entails B if and only if A is inconsistent with not –B and that when this relation holds, the truth of B is deducible from the truth of A. Because we can never be presented with the fact that would make A true in the first place. What OC goes on to say here is that since A and B are statements, we know that events might occur that would confirm them and also that events might occur that would disconfirm them. And we describe the entailment of B by A by saying that if A is true, then if it is put to the test it will be confirmed and as a result B will also, inescapably, be tested and confirmed to at least the same extent and if B is tested and disconfirmed, A will be tested and disconfirmed to at least equal extent. This in rough terms is the method OC follows in establishing entailment relations between two uncombined statements. In the following pages, and especially in chap. 2, the method is refined so as to bring within its competence clusters of more than two statements. It is a method that leads to a system of deduction which mirrors many intuitions of long standing about the logic of conditionals. And the system is one that enables us to determine the entailment relations between statements by reference to events, which may occur and be recorded, and not to relations between things, which are of necessity beyond direct detection.

It will be helpful, at this juncture, to indicate in a general way the structure an inconsistency must have under this proposal. I will assume for the purposes of these comments, that the inconsistencies of which I speak are ones all of whose components are relevent. Generally, any conjunctive expression that contains an inconsistency is itself inconsistent. Thus, 'p.–p.q.r' is inconsistent, as is 'p.–p'. These remarks, however, will be addressed only to those expressions, such as 'p.–p' or 'If p then q and p and –q', in which each part is essential to the existence of a contradiction.

In the first place, if one component of an inconsistent conjunction is provable, then its proof must disprove the rest of the conjunction. Suppose we have a conjunction A.B.C and that C is provable but that its proof does not disprove the conjunction of A and B. Then if nature presents us with the proof of C, without accompanying this with the evidence that would disprove A.B, it seems that we may consistently maintain that A and B are true, notwithstanding that we have conclusive evidence of the truth of C, and we are not in a position to claim that the conjunction A.B.C is logically inconsistent. For example, taking a formula with two components only, in the conjunction 'It is possible that something is X and also possible that something is –X,' the proof of the first component, by instantiation of X, confirms the denial of the second, but it does not disprove the second. It is logically possible that both components are true, and therefore the conjunction is not inconsistent.

A second feature of an inconsistent conjunction is that it is true of each of its components that the denial of the component is confirmed, to the extent that it can be confirmed, by the attempt to confirm the rest of the conjunction. This stipulation is made necessary by the fact that the negation of the disconfirmation of a conditional is not equivalent to the confirmation of the conditional. The point is illustrated by the problem of whether '–p' entails 'If p then q'. This will be so if

'–p.–(If p then q)'

is inconsistent. Now '–(If p then q)' is confirmed by 'p.–q,' which truth functionally (hereafter TF) implies 'p', the disconfirmation of the antecedent '–p.' Thus, if our rules require of an inconsistency only that one of its components be disconfirmed by the confirmation of the rest of the conjunction, '–p.–(If p then q)' would be inconsistent. But '–p,' though inconsistent with 'p.–q', does not TF imply either of the confirmations of 'If p then q', and so '–p.–(If p then q)' is not inconsistent, and 'p' does not entail 'If p then q'. Otherwise put, inconsistency does not obtain in the conjunction '–p.–(If p then q)' because it is possible for events to afford us confirmation of the component '–p' without confirming the denial of the remaining component to its fullest extent.

Special attention to each component is not necessary when each member of a conjunction is elementary. For such expressions, the negation of the disconfirmation of a component is equivalent to the confirmation of its assertion, and if one component has the appropriate characteristic, the other has it too. Thus if we know that each of A and B is elementary, then if

A TF implies –B

we know that

B TF implies –A.

As I have observed, this straightforward arrangement is destroyed by the introduction of conditionals.

In short, a conjunction is inconsistent if and only if (1) the proof of one of its components disproves the rest of the conjunction and (2) the denial of each component is confirmed, to the extent possible, by the confirmation of the rest of the conjunction.

Let us see how these more focused considerations bear on the question, raised earlier, of whether 'If p then q' entails 'If –q then –p'. As was said, this will be the case if

'If p then q and –(If –q then –p)'

is inconsistent. It is easy to show that this conjunction satisfies both of these conditions on inconsistency. In reviewing the effect of the confirmation or disconfirmation of one component on another, we need not consider that either confirmation or disconfirmation is fulfilled in the actual world. We need only ask what would be the effect of confirming or disconfirming events were they to be fulfilled. With regard to the first requirement on inconsistency, '–(If –q then –p)' is the only provable component of the mentioned conjunction, and its proof, '(p.–q)', disproves 'If p then q'. As for the second requirement, we see that if nature furnishes us with confirmation, in two phases or segments, (i) and (ii), of the occasion to which 'If p then q' applies, we have, attaching the indication of phase to each confirmation form, '(i)p.q' and '(ii)–p.–q'. These are the alternative confirmations of 'If p then q' and 'If –q then –p'. The argument may be presented schematically.

> To be demonstrated: that the confirmations of each of 'If p then q' and '–(If –q then –p)' truth functionally imply the confirmations of the denial of the other.

Confirmation during	*Component*	*Denial of the other*
	'If p the q'	'If –q then –p'
Segment (i)	'(i)p.q'	'(i)p.q'
Segment (ii)	'(ii)–p.–q'	'(ii)–p.–q'

We see here that by the time 'If p then q' is confirmed in each alternative way, 'If –q then –p' also receives confirmation in each alternative way. Also, confirmation of '–(If –q then –p), by 'p.–q', TF implies the confirmation of the denial of 'If p then q' as follows:

Confirmation during	*Component*	*Denial of the other*
	'–(If –q then –p)'	'–(If p then q)'
Any segment	'p.–q'	'p.–q'

Confirmation of '–(If –q then –p)' and '–(If p then q)' by 'p.–q' may occur during any segment because it is conclusive and renders happenings in other segments immaterial. It is therefore unnecessary to indicate the phase during which the proof occurs. By our preliminary criterion, then, 'If p then q' is inconsistent with '–(If –q then –p)' and thus 'If p then q' entails 'If –q then –p'.

V

The view just outlined is substantially different from the theory developed by R. Stalnaker and David Lewis[3] and that has come to assume an important position in discussions of the conditional. According to Stalnaker's version of this theory, a conclusion about the truth or falsity of a conditional is to be arrived at in the following way. 'First, add the antecedent [the 'condition' of OC] (hypothetically) to your stock of beliefs; second, make whatever adjustments are required to maintain consistency (without modifying the hypothetical belief in the antecedent); finally, consider whether or not the consequent is then true.'[4]

This operation, however, produces a state of hypothetical belief, not the truth conditions which (Stalnaker says) are needed for a theory of conditionals. But, the theory goes on, belief conditions can be translated into truth conditions by transposing the 'hypothetical beliefs' into their 'ontological analogues', possible worlds. If the antecedent is not true, one thinks of it as fulfilled in a possible world that differs from our world only in that it contains the antecedent and such other changed items as may be necessary for its accommodation. 'If p then q' becomes true if *p* is present in no possible world or if *q* obtains in the closest accessible world containing *p*. The theory thus proposes that we give up the attempt to find truth conditions for conditionals exclusively among the elementary statements true or false of this world. When our world lacks either disconfirmation of the conditional in question or the form of confirmation represented by the truth of antecedent and consequent, we are to seek support or infirmation among the contents of possible worlds. The influence of the form of atomism to which I have referred is at work here. Singular statements are to be reduced to statements asserting occurrences of events and truth functional compounds of these. If conditionals cannot be reduced to occurrence statements asserted of things in the actual world, it is necessary to assign the required occurrences to possible worlds.

The theory contains assumptions that make the logical relations of conditionals different from those that hold among the conditionals examined in OC. One is that the truth condition of a conditional is identified with the fulfillment of events in combination, specifically the joint occurrence of the events referred to in the 'if' clause and the 'then' clause. This obliges us, as we have seen, when the requisite events do not occur in the actual world, to say that if the conditional is true, the events that constitute its truth condition must be occurring in some 'possible world'. There is thus an awareness in possible worlds theory of a fact that is also recognized in OC: that in many cases, evidence for the truth or falsity of a conditional is in principle unavailable. But in OC, the truth of a conditional is not identified with actual or possible occurrences of events. For OC, because the truth condition of a conditional does not consist in the occurrence of an event or events, it is not possible to organize the logic of conditionals on the basis of their truth conditions. We can have no direct access to the truth conditions of conditionals and so, if we had to determine the logical relations of conditionals by reference to their truth conditions, the logic of conditionals would have no application. The theory of OC, with regard to the conditional, is that to establish its logical links with other statements, we need not know, indeed cannot know, how the fulfillment of its objective state, the situation that would make it true, might be ascertained. We need only know how, in principle, the conditional might be confirmed or disconfirmed. If we know this we can say, within the limits of anything that might offer itself for observation, what the consequences of confirmation or disconfirmation would be for the confirmation or disconfirmation of other statements. This is why the issue of the structure of support and infirmation is crucial to the logic of statements in OC.

And this problem of the structure of confirmation directs us to a second assumption of possible worlds theory—the assumption that not only is confirmation (in the actual or a possible world) identical with fulfillment of the truth conditions, but that this confirmation must consist in the occurrence, in some world, of the events mentioned by the 'if' clause and the 'then' clause. This has the consequence that the conditional relation holds between all occurrent events. According to Stalnaker[5] if 'p' is true in our world, then our world is the closest accessible world containing *p* and, because the conditional is proved by its confirmation, the conditional relation holds between *p* and all this world's happenings. If this stone is on the ground and the sun shines, then we are bound to accept both 'If the sun shines, this stone is on the ground' and 'If this stone is on the ground, then the sun shines'. But if one uses the conditional in the way of OC, to claim that fulfillment of the condition is sufficient to ensure fulfillment of the conclusion, this is just the sort of result he is justified in resisting.

The strangeness of the result is recognized by Lewis who writes[6] that it seems odd to be able to 'pick *two completely unrelated* truths ϕ and ψ and, on the strength of their truth . . . to assert the counterfactual $\phi \Box\!\!\rightarrow \mu$'. Lewis, however, defends the view that 'If p then q' is proved by the truth of 'p' and 'q' by saying' 'But oddity is not falsity; not everything true is a good thing to say. In fact the oddity dazzles us. It blinds us to the truth value of the sentences, and we can make no confident judgment one way or the other'.

It seems to me that what dazzles us, and prevents us from seeing the obvious, is the language of possible worlds. It is just because we know that any arbitrary *p* and *q* may be 'completely unrelated' that we refuse to deduce the truth of 'If p then q' from the joint truth of 'p' and 'q'. In fact, most concurrent events are unrelated to each other, whereas the relatedness of *p* and *q* is what 'If p then q' is supposed to express. The joint occurrence of *p* and *q*, in whatever world, cannot of itself stand as the truth condition of 'If p then q'.

The attempt to manage the logic of conditionals with possible worlds semantics has other results that ensure that the conditionals of this theory have logical characteristics that are in conflict with those the conditional has traditionally been thought to possess and that the conditional I have called 'ordinary' does possess in OC. Two prominent consequences of the possible worlds approach, now widely accepted, are that contraposition and transitivity of conditionals become invalid. I shall argue that these results are wrong, at least to the extent that they might be taken to apply to the type of conditional addressed in OC. The question of transitivity is examined at pp. 71–74, and an argument of Frank Jackson's against contraposition is discussed in section 8. I review here Stalnaker's original attack on contraposition based on considerations of possible worlds. The defeat of this relation in possible worlds theory comes about because once a conditional is supplied with the recommended form, 'If p then q' is no longer the contrapositive, in the material sense, of 'If –q then –p'. The first becomes 'p is true in a possible world · ⊃ · q is true in the closest world containing p', and the second becomes '–q is true in a possible world · ⊃ · –p is true in the closest world containing –q.[8]' Obviously, the condition of the second translation is not equivalent to the denial of the conclusion of the first translation. Stalnaker defends this consequence by arguing that one might accept 'If p then q' as true while rejecting as false 'If –q then –p'. This would be the situation if 'q' is true in the closest world of which 'p' is true and 'p' is true in the different world that is the closest world in which '–q' is true.

Stalnaker illustrates the theory with the sentence (X), 'If the U.S. halts the bombing, then North Vietnam will not agree to negotiate' and says (p. 174), 'A person would believe that this statement is true if he thought that the North Vietnamese were determined to press for complete withdrawal of U.S. troops. But he surely would deny the contrapositive, "If North Vietnam agrees to negotiate, then the U.S. will not have halted the bombing."' Yet if (X) were a true conditional, undoubtedly one *must* conclude that if North Vietnam agrees to negotiate, the U.S. will not have stopped the bombing, because if the U.S. halted the bombing and North Vietnam agreed to negotiate, (X), although true by hypothesis, would be disconfirmed. On the same page, indeed, Stalnaker says that although contraposition fails, '*Modus tollens* is valid for the conditional: From A > –B, and B, one can infer –A' (where '>' is Stalnaker's sign for the conditional relation). But when the denial of contraposition and acceptance of *modus tollens* are both applied to (X), it must be the case (1) that if (X) is true it does not follow that if North Vietnam agrees to negotiate, the U.S. will not have stopped the bombing but (2) that, again if (X) is true, it does follow that if it is also true that North Vietnam agrees to negotiate, the U.S. will not have stopped the bombing. It is hard to think that these two results are not in conflict.

Before I attempt to unravel this problem, let me point to a feature of Stalnaker's argument that I think makes the issues more difficult to see. I refer to a characteristic of statement (X) itself. If it is true, it is true only if treated as an 'even if' conditional; however, in that case neither contraposition nor *modus tollens* applies to it. The point of the 'even if' conditional is to assert that the occurrence of the condition is not sufficient to prevent the fulfillment of the conclusion. And this is different from the claim that the fulfillment of the condition is sufficient to ensure the fulfillment of the conclusion. To say, 'Even if I jump on the floor the house will remain standing' is not to say that jumping on the floor is sufficient to ensure that the house will remain standing. Stalnaker says that a person would believe (X) if he thought that the North Vietnamese would be satisfied with nothing less than the complete withdrawal of U.S. troops. But if this is so, then it would not be a cessation of the bombing that would entitle us to conclude that the North Vietnamese will not agree to negotiate. What (X) says is that a cessation of the bombing will not of itself affect the situation one way or another. As Stalnaker himself points out, if North Vietnam does agree to negotiate, one cannot conclude that the U.S. will not have stopped the bombing; and I suggest that this is so whether the remark is taken as a rejection of contraposition or of *modus tollens* and that this is because (X) is not a proper conditional. Whatever may be the precise characteristics of the logic of the 'even if' conditional, they are different from those of the conditional I have called 'ordinary'. (See my comments on 'even if' conditionals in chap. 2 at pp. 54–55.) By lumping together 'even if' conditionals and those that assert that fulfillment of the condition is sufficient to ensure the truth of the conclusion, Stalnaker has obscured the difference between the two types of statement. One cannot conclude that, because contraposition does not hold for 'even if' conditionals, it does not hold for conditionals of other types.[9] I will argue that it is unreasonable to suppose that contraposition does not apply to ordinary conditionals.

Because the conditional about North Vietnam probably does not qualify as a conditional of the sort that says that the fulfillment of the condition justifies, in an empirical sense, the assertion of the conclusion, it is appropriate to consider one that is more clearly of this type—

(1) 'If the vase is dropped on the floor, it will break.'

—and ask again how Stalnaker's possible worlds logician can hold that while the form,

(2) 'If p then q.p.–q'

is inconsistent, the form

(3) 'If p then q and –(If –q then –p)'

is not. The answer seems to be that (2) asserts 'p' and therefore the actual world is the closest world in which 'p' is true and so, by 'If p then q', 'q' must be true. But 'q' is asserted to be false and thus (2) is inconsistent. But in (3), neither 'p' nor '–q' is asserted to be true in the actual world and so it is possible that 'p' is true in the closest world in which '–q' is true. Although this position is coherent, I suggest that it does not represent a conception upon which the regular use of conditionals is based. In our dealings with the world, we accumulate knowledge and opinions about the organization of events in the actual world. We want to have reliable understandings about what, under particular circumstances, a situation can be expected to reveal in the way of further events and characteristics. Conditionals, in their most central employment, are about present or prospective happenings in the actual world and the conclusions one may draw upon the occurrence of these happenings. In helping us to impose order on the world, they help us to determine our own actions. Consequently, we are not concerned to know, when 'p' is false, whether 'q' is true in the closest (but by definition unactual) world in which 'p' is true. (How could one find out such a thing?) We want to know what would happen, what conclusion one might draw, if the event expressed by 'p' is actualized by its occurrence. The possible worlds logician will not permit us to infer from (1) the truth of

(4) 'If the vase is unbroken it has not been dropped'

because the nearest world in which the vase is unbroken may be different from the nearest world in which the vase is dropped. Yet suppose we regard (1) and (4) as being about the occurrence or otherwise of events in the observable world and understand 'occurrence in the observable world' as something that may be happening now or that may occur in the near future. Then (1) and (4) may be taken as saying,

(1)(i) 'If the vase is dropped in the observable world, it will break' and
(4)(i) 'If the vase is unbroken in the observable world, it has not been dropped.'

I expect that even the possible worlds logician will agree that (4)(i) can be inferred from (1)(i). Because if one were to find the vase unbroken in the observable world, then it has not been dropped or (1)(i) is false. I suspect also that (1)(i) and (4)(i) do express the meanings of (1) and (4) in normal speech and that this is why contraposition traditionally has been thought to hold for conditionals. Although a conditional about prospective occurrences does not assume the fulfillment of the affirmation or denial of its 'if' clause or its 'then' clause, it does assume that such fulfillments as do occur will occur in the observable world. If this is so, then it is not surprising that it is the common view of conditionals that they entail their own contrapositives.[10]

It is worth observing also that the possible worlds logician, no less than the rest of us, must estimate the truth of an unfulfilled conditional on the basis of his knowledge of the experienced world. He is not in a position to take a privileged peek at the contents of possible worlds but must judge a conditional's truth or falsity in the light of his knowledge of this world's happenings. And if his judgment is grounded on the best type of evidence—evidence of a sort that supports the truth of a statement that itself would necessitate the truth of the unfulfilled conditional—there is every reason to suppose that if he decides, on the evidence, that 'If p then q' is true, the evidence on which he relies will support 'If –q then –p' to an equal extent. Consider how, as possible worlds logicians, we would approach the question of the truth or falsity of the unfulfilled conditional, 'If p then q', of the type that expresses the claim that the fulfillment of the condition justifies the conclusion that the 'then' clause is true. We ask whether, in the closest accessible world in which the conditioning event is fulfilled, the projected event is also fulfilled. But because no conclusion on this issue can have direct observable support, we are forced to decide, before furnishing possible worlds with appropriate events, whether the conditional is true or false. We then, like everyone else, search for indirect evidence for the truth of the conditional. And the sole form of indirect support that necessitates the singular conditional will be a true universal conditional from which the singular conditional may be deduced. (I show in chap. 3 that a universal conditional entails its corresponding singular form.) But if we are to accept a general conditional as true, it must have been actually confirmed itself, and it turns out that the support and infirmation of a universal conditional are just the same as the support and infirmation of the contrapositive of the conditional.

Let us suppose that the conditional in question is, 'If interest rates reach bottom in July, banker's acceptance will be below prime'. How will the possible worlds logician justify this conditional if interest rates do not in fact reach bottom in July? He would have strong indirect support for this assertion if it were true that there is ample evidence for the truth of 'Whenever interest rates reach the bottom of a cycle, banker's acceptance is below prime'. But what does the evidence for this claim consist in? We must satisfy ourselves that in all known cases banker's acceptance is below prime at times when interest rates are at the low point of a cycle and also that whenever banker's acceptance is equal to or above prime, the interest rate cycle has not reached bottom; otherwise the general statement would be falsified. But this is precisely the procedure we would follow in attempting to establish the truth of 'Whenever banker's acceptance is not below prime, the interest rate cycle has not reached bottom', which, if true, affords indirect support for 'If banker's acceptance is not below prime in July, interest rates will not have reached bottom', which is as strong as the support furnished to the original conditional by *its* general version. Thus the evidence confirming a general statement that entails a singular conditional confirms to the same extent a general statement that entails the contrapositive of the conditional. In addition, the conditional and its contrapositive will each be disconfirmed and disproved if it should happen that banker's acceptance is not below prime but interest rates do reach bottom in July. Thus we have no justification for attributing presences and absences of *p* and *q* to possible worlds in such a way as to make 'If p then q' true and 'If –q then –p' false. And the evidence we may have that makes plausible the assignment of *q* to the closest world containing *p* also makes plausible, to the same degree, the assignment of –*p* to the closest world containing –*q*.

Factual statements are confirmed or disconfirmed by occurrences of events in the actual world. It is the object of OC to present the logical relations of statements on the basis of the relations they bear to the events that would support or undermine them. Because actual events are the only ones that can assist us in justifying our estimates of the truth or falsity of conditionals, we should consider the logic of conditionals in the light of events which, if they occurred in the actual world, would provide the conditional with its direct support and its direct disconfirmation. According to OC, if A is to entail B, the confirmation of A must support B with at least equal strength and the disconfirmation of B must disconfirm A with at least equal strength. In chapter 2, it is shown that an appropriate generalization of this notion makes it possible to apply it to complex forms.

In OC, to say that the event or situation A is possible is to say that it may occur in the actual world, not that it is occurring in a possible but unactual world. To the state projected by the latter assertion we lack all empirical access; the former assertion, as we will see in chap. 3 (p. 88), is provable by an occurrence of A and it is disconfirmable, although not disprovable, by a statement that expresses a conjunction of assertions about actual states that are not occurrences of A.

I do not discuss the recent theories that identify the singular conditional with a form of probability statement. Although the study of statements of conditional probability is an important undertaking in itself, I doubt that the treatment of the singular conditional as a statement of conditional probability can be relied upon as a guide to the logic of the former. In general terms, the statement, 'It is probable that p' is logically independent of 'p', and the extent to which 'If p then q' and 'q is probable given p' are logically connected is by no means clear. And it will not become clear, I would submit, until we have a firm grasp of the logical properties of each. To say, for example, that 'If p then q' does not entail its contrapositive on the ground that 'q is probable, given p' does not entail *its* contrapositive, prejudges the question of the logic of 'If p then q'. The futility of equating conditionality with conditional probability without first addressing the question of the logic of conditionals is illustrated by Ernest Adams's approach to contraposition. He says[11] that if all of A lies in B but a large part of B lies outside A, then we may accept 'If it's in B, then it's not in A' as true because it is highly probable that things in B are not in A. But the probability that an A is outside B is zero and so, according to Adams, contraposition is invalid. But on the described facts, it is not true that if a thing is in B, this ensures that it is not in A; and if it *were* true it would also be true that if a thing were within A, it would not be within B. Adams begs the entire question of the logic of conditionals that are not statements of probability. The essay confines itself to the problems of the logic of singular and general conditionals and their denials and I make no attempt to address the issues relating to conditional probabilities.

VI

If the logical relations between assertions are to be assessed in terms of the associations between statements that supply their support and infirmation, it is important that we examine with care the formation, and in particular the limitations on the formation, of confirmations and disconfirmations of the original or 'parent' statements.

A statement is something that identifies one or more subjects and says something about them in a way that permits us to collect evidence for or against its truth. An expression cannot fulfill the requirements of statementhood, therefore, if its confirmation or disconfirmation is a contradiction. A statement says that an objective state obtains. If an evidential statement is *a priori* bound to be contradictory, it cannot either support or infirm the existence of the objective state and we conclude that the expression for which the assertion was to act as support or infirmation is not a statement and not capable of being true or false. Thus, when one comes to consider the application of logical principles to discourse, he must think of the confirmations and disconfirmations of statements as being formally contingent. For this reason, expressions having contradictory confirmations or disconfirmations are not well formed and may not take positions in entailments.

It follows from this that a conditional cannot be an entailment in OC: if 'p entails q' is true, then 'If p then q' is not well formed. This is not to deny that in everyday use, 'p entails q' is often naturally expressed by the form 'If p then q'. It is just that if we are to reserve the 'If–then' form for the expression of ordinary conditionality, we may not use this form for the assertion of entailment statements. If 'p' does entail 'q', then 'p' is inconsistent with 'not–q' and its confirmation is inconsistent with the confirmation of 'not–q'. But then the form 'p and not–q' could not serve as a disconfirmation of 'If p then q' because disconfirmations as well as confirmations must be contingent. (For specifications of conditionals, see pp. 61–62, and for a discussion of the differences between conditionals and entailments statements, see chap. 6, sec. 4.)

VII

Confirmations and disconfirmations, then, must be formed with assertions that are logically contingent; if they are not, the expressions they purport to confirm or disconfirm will not have entailment relations. But this is not the only constraint on the statements taking part in deductive transformations. Unspoken assumptions are also usually vital to the validity of deductions. For example, similar referring expressions must refer to the same thing throughout the deduction. Otherwise, an interpretation of form A appearing on the left side of the entailment would make a statement different from the one that would be made by an interpretation of A appearing on the right side. The law that A is deducible from A applies to singular statements on the assumption that the object or thing identified by the second A is the same as that identified by the first A. 'Smith is blue eyed' does not entail 'Smith is blue eyed' if the person identified by the first 'Smith' is my neighbor and the person referred to by the second 'Smith' is a man who lives farther down the block. It is then a factual possibility, but not a logical truth, that their eyes have the same color.

Recognition of the role of unexpressed assumptions is of special importance when we come to deal with the logical relations of conditionals. The problem arises in this way. Many of the conditions on which the truth values of ordinary conditionals rest are implicit. We allow that 'If I release this pencil, it will fall' is true; and yet we know that it is not, on its face, true without qualification. If I ignore the context in which the statement is made, I can readily demonstrate its falsity by holding the pencil against my palm with my other hand, by sticking it to my palm with glue, or by releasing it over a strong updraft. But we accept the statement as true because we jointly assume that other relevant conditions are satisfied. This is a familiar feature of our use of conditionals. Speech would be painfully strenuous if we insisted that all the known relevant conditions must be made explicit in all our conditional statements. But I suggest that if we permit conditionals to be true on the basis of the satisfaction of assumed conditions, we must regard these assumed conditions as holding throughout the transformations in which they participate. We must observe the contextual continuity of the assumed relevant conditions throughout the deduction. If we do not respect this constraint, distortion inevitably results.

The confusion that may occur when one ignores the requirement of contextual continuity of relevant assumptions is easy to illustrate. For example, the theory of conditionals based on the material relation holds that 'If p then q' entails 'If p and r then q', on the ground that the disconfirmation of the latter, 'p.r.–q'. contains the disconfirmation of the former. But, Stalnaker has said, '[W]e cannot always strengthen the antecedent of a true conditional and have it remain true. Consider, "If this match were struck, it would light," and "If this match had been soaked in water overnight *and* it were struck, it would light"'[12]. Stalnaker concludes that 'If p then q' does not entail 'If p and r than q'—that although truth is preserved through the strengthening of the condition of a material conditional, this cannot be relied upon to happen in the case of the standard or subjunctive conditional. But it seems to me that we do not really have here a counterinstance of the relation in question. We assert the truth of 'If this match were struck, it would light' on the assumption that at the time the match is struck, if it is struck, it will be dry. But the condition clause of 'If this match had been soaked in water overnight *and* it were struck, it would light' makes it plain that if *its* condition is fulfilled, the assumption on the basis of which the first statement was asserted will be false. Thus, the second statement involves the rejection of one of the relevant assumptions of the first statement, and the relation is not an application of the putative law. This is seen if we remember that the full statement of the abbreviated conditional, 'If this match were struck, it would light' would be something like, 'If this match were struck, and if it has not been soaked in water, and if, . . . then it would light'. This statement does not entail, even when antecedent and consequent are considered as material conditionals, 'If this match had been soaked in water overnight *and* it were struck, it would light', and it is not to be expected that the falsity of the latter is inconsistent with the truth of the former.

Frank Jackson has objected to attempts to preserve the validity of the deduction of 'If p and r then q' from 'If p then q' by making explicit the assumptions present in particular applications of 'If p then q'. He says that assumptions may be added indefinitely and that if this is permitted the statement becomes like a 'dog chasing its tail'.[13] But in discourse conducted in good faith, assumed elements taken to be present in the context of application are limited in number and understood by the participants. In any case I suggest that the assumption of the dryness of the match in Stalnaker's example is vital to its apparent success. In accepting the truth of 'If this match were struck, it would light', we must assume that the match referred to is dry.

Demonstration by example in logic is invaluable. It is the only method by which we can show that an abstract theory fails to reflect the logical movements of the language we speak. But a demonstration by this method that a logical law is invalid is nullified if it relies for its success on changes in the context upon which the truth of the antecedent depends. What logical laws seek to express are formal relations not merely between sentences but between statements; and statements are always made in context. The law of the propositional calculus that

$$((-p \lor q)(rp)) \supset (q)$$

would not survive the sort of analysis to which Stalnaker subjects the conditional. Let 'p' be 'This match is struck', 'q' 'This match lights' and 'r' 'This match is wet'. Would we be allowed to say that the law is invalid because 'This match is not struck or it lights' and 'This match is wet and it is struck' may be true although 'This match lights' is false? I think not. One would defend the law by saying that the truth of 'This match is not struck or it lights', in the context of application, assumes that the match has not been placed in water. Thus a fuller spelling out of 'This match is not struck or it lights' would be, 'This match is not struck or it is wet or it lights' and $((-p \lor r \lor q)\ (rp) \supset (q)$ is not a law of the propositional calculus.

We therefore do not have an instance of a true antecedent and false consequent, which would be required to invalidate the mentioned law, $((-p \vee q)(rp)) \supset (q)$.

I have reviewed the question of contextual continuity because it is one that is frequently ignored in discussions of conditionals. Its neglect makes possible, for example, a certain argument against the standard view that conditionals are transitive. I show in chapter 2 that this argument fails once the requirement of the continuity of assumptions is taken into account (see pp. 72–74). I have mentioned Stalnaker's example in particular because his argument gives the wrong reason for an important truth. 'If p then q' does not entail 'If p and r then q'. It is shown in chap. 2 that we cannot conclude, from the truth of 'If p then q' that 'r' is a relevant part of the condition for the occurrence of 'q' and that it is for this reason that the entailment fails. On similar grounds, the present theory also rejects the entailment of 'If p then q v r' by 'If p then q'. And we will see in chap. 5 that this result has important repercussions for the theory of confirmation.

Only a small part of the environmental background of a statement is included in its assumed relevant context. In the case of the match, we know that the surface on which it is struck must be dry, sufficient oxygen present, the match not wet, and so on. The presence or absence of an indefinite number of other factors is irrelevant. Discourse succeeds in part because its participants have a common understanding of the contexts in which statements are made. If this understanding is not present, then communication breaks down. The circumstances under which an assertion is justified or placed in doubt will then be different for different speakers.

VIII

I have maintained that evidential statements must not be contradictory or tautologous, and I have argued that they must not be contrary to the contextual assumptions upon which their parent statements are based. A third stipulation is that they must not contravene the rules that govern what I call their 'conventional' relations with other expressions. To see this it is necessary to reflect for a moment on the functions of rules of language.

The existence of a language of fact is made possible by a system of conventions, a domain of language that concerns the prescribed and permitted uses of expressions that receive application in the language of fact. This is so for obvious reasons. One could not tell, or find out, whether a statement of fact is true in the absence of standard ways, accepted by other users of the language, in which the expressions contained in the statement are related to the experienced world. The significant use of the sentence 'This leaf is green' in a particular statement depends upon the existence of standard ways of applying the identifying expression 'this leaf' and the predicate 'is green'. Language rules determine the content of the confirmation statements; therefore without a grasp of these rules one could not map the relations of statements to each other.

Rules act within factual speech implicitly. In talking about happenings and the qualities of things, one does not talk about rules. But before he can speak in a language, a person must know some of its rules and it must be possible therefore not only to use rules in a language but also to give expression to them, to instruct others in their use.

Rules of this kind are often called rules of meaning, in the sense of 'meaning' in which to know the meaning of an expression is to know how it is to be used in intelligible discourse. In this broad acceptation, rules of meaning include lexical rules, grammatical regulations, and logical specifications and laws. Let us separate out from the many different types of linguistic prescription certain of those rules that are arbitrary or 'conventional' in that although binding on discourse once adopted, they need not have been the rules we chose to adopt. And let us attend in particular to three types of such rule, all of which are arbitrary in this sense and concern the relations we impose between linguistic expressions and the world. (1) The first type is called the 'ostensive' rule and it links words and expressions to external events by associating them with sounds, sights, feelings, touches, tastes, and smells. The use of these rules can only be explained by the association, with the aid of gestures and pointing, of expressions with nonlinguistic experience. (2) The second type of rule, the 'testing' rule, makes the question of whether a thing or event has a given characteristic turn on the outcome of an experiment. The question, for example, of whether a substance is a base turns on whether, when mixed with acid, it forms salt and water. Understanding how to apply testing rules necessitates that the user already understands many ostensive rules. (3) The third type of rule will be called the 'convention' and it concerns the prescribed connections between expressions, especially predicates, and other expressions. The bindingness of conventions stems from the fact of the adoption of ostensive rules. Red and blue having the names they have, 'is blue' and 'is red' cannot both be true of the same patch of color. When one understands how to characterize a thing as opaque on the ground that one cannot see through it and as transparent if one can see through it, he is possessed of the conventional knowledge that if a thing is opaque then it is not transparent. Because conventions arise in this indirect way, they will sometimes be referred to as 'derived rules.' It must not be thought that conventions are mere matters of linguistic etiquette. They are not expressions of preference for 'isn't' over 'ain't' or for 'shall,' in appropriate circumstances, over 'will.' They are parts of the normative fabric of language and the abuse of them reduces speech to gabble.

In natural language, logical laws can be relied upon to apply normally only to sentence forms that are conventionally independent of each other. Consider, for example, the law of the propositional calculus that,

((–pv –q) (q)) entails (–p)

and its application where 'p' stands for 'It snows' and 'q' for 'It snows heavily'. The law requires us to say that from 'Either it does not snow or it does not snow heavily' and 'It snows heavily' one deduces, absurdly, 'It does not snow'. Thus, the *modus tollens* law of the propositional calculus, of which this is a version, may be destroyed by applications that ignore conventional relations between the predicates of the interpreting statements. And regard for conventional relations is as important to OC as it is for the propositional calculus. In OC, conventional rules are a governing factor in the formation of confirmations and disconfirmations. And if the logical connections between statements are in turn controlled by the relations statements have to their confirmations and disconfirmations, it follows that the laws of OC concerning the conditional may be effectively distorted where the predicates of the 'if' and 'then' clauses are related by convention. This is why I drew attention, in the Preface, to the 'suppositive' conditional in which there is a conventional connection between the 'if' clause and the 'then' clause, and distinguished this type of conditional from the ordinary one. If the suppositive conditional were accepted as ordinary, the OC position that contraposition of the conditional is valid would properly be called in question: it would not make sense to say that, 'If it snows, then it does not snow heavily' entails 'If it snows heavily, then it does not snow'. It is this sort of example that Frank Jackson employs to justify his denial of the validity of contraposition in *Conditionals*.[14] But the pursuit of this kind of analysis, which is deaf to the semantic relations between the predicates involved, would destroy *modus tollens* for conditionals, even, as just noticed, for material conditionals.

No logic of factual language will be complete that does not expose the effect of rules on the structure of support and disconfirmation and also on the logical relations of rule statements to each other. The subject matter of chapter 4 will include both these topics. But once we have brought to the surface the role of conventions in the formation of factual language, we are also compelled to recognize a different influence on discourse, one that concerns the limits of what we can say and still make sense. The assertions that govern these limits—I shall call them 'conceptual necessities'—are different from conventions in that their observance is unavoidable. They cannot be changed by common consent as can conventions; the consequences of their violation even by all speakers would be to make language unintelligible.

It will be the principal object of chapter 4, however, to set forth a logic of conventions in which the convention is understood to be a type of conditional. It will be one of the aims of that chapter to show that a logic of conventions can be constructed in a way that parallels that of the factual conditional. The concept of entailment employed in chapter 4 to bring to light the logical relations between language rules is an extension of the entailment concept to be developed in chapter 2. In chapter 6, I turn to the question of the logical relations of a certain type of conceptual necessity—the logical statement.

IX

At this point in the journey it is possible to consider the distance we have come and to see in profile the terrain that lies ahead. It appears that conditionals belong to the class of nonelementary statements and are not proved by the direct evidence in their support. To assert a conditional, for example, 'If p then q', is to say that reality is patterned in such a way, things are so related, that the truth of 'q' would be assured by the truth of 'p', but not that 'If p then q' is reducible to any combination of the truths of the direct evidence statements that would confirm it. There is an unlimited number of potential 'conditions' the occurrence of which might ensure the truth of 'q', and one cannot infer, from the truth simply of 'p' and 'q', that it is the truth of 'p' that is sufficient to secure the truth of 'q'. The logical relations of 'If p then q' may be appraised, however, in the light of the relations it bears to the forms of its confirmation and disconfirmation, and our first task will be to describe the relations between factual conditionals and the evidential forms that directly support and infirm them and also to show how the logical connections between conditionals and between conditionals and nonconditionals, can be charted by reference to the truth functional relations of the forms that provide their fulfillments and rejections. This will be done for singular factual conditionals in chapter 2 and for general factual conditionals in chapter 3.

The theory of deduction that regulates the relations of formal inference among statements in OC is based on the familiar notion that, when a conjunction of statements is inconsistent, the denial of any part of the conjunction is justified by the truth of the other part. Thus if,

(A and B and C)

is inconsistent (not –A) is inferable from (B and C) and (not –B or not –C) is inferable from (A).

For the purposes of OC there are two qualifications on this rule. The first is that nothing is to be inferred from an inconsistency. Thus (not–A) may not be deduced from (B and C) if (B and C) is itself formally inconsistent. We shall also see (p. 159) that as a result of this qualification, a tautology (the denial of an inconsistency) may not be deduced from any statement. It is not suggested that this is a necessary feature of any proper theory of deduction, but it makes it possible to say that the relation of formal deduction does not obtain unless the truth of the antecedent justifies the inference that the consequent is true in the way described in the chapters that follow. The coherence of the position is argued for in section 2 of chapter 6.

The second qualification is that in OC, for reasons I have explained, the formal inconsistency of a conjunction is identified not directly in terms of the truth or falsity of its sentential parts, but on the basis of the relations these parts have to their direct support and infirmation, because it is to the support and infirmation of a statement that, in principle, we always have access. (A and B) is inconsistent when it can be said that, if A is true, testing it will result not only in its own confirmation, but also in that of not–B to at least equal extent and the same can be said of B with respect to not–A.

In expressing the claim that one statement is logically inferable from, or is entailed by, another, we often in informal speech use the 'if-then' locution. Instead of saying, for example, that the conjunction of 'p v q' and '–p' entails 'q', we as often say that if 'p v q' then if '–p' then 'q'. But this mode of representation, while natural sounding, and familiar in ordinary discourse, leads us to think of the entailment statement as a form of conditional, which is a mistake. In general, the truth of (A entails B) is founded on the formal inconsistency of (A and not–B), whereas if A is inconsistent with not–B, the conditional (If A then B) is not well formed because its disconfirmation is an inconsistency. Thus I will take care to use such expressions as 'is entailed by', 'is deducible from', 'is inferable from' to signify the entailment relation and reserve the 'if–then' form for the presentation of conditional relations. It will then be easier to keep the formal relations of expressions separate from conditional ones.

In saying that if 'p v q' then if '–p' then 'q', we are tempted to suppose that the deductive entitlement arises at the first 'then' as well as at the second. But it is not hard to see that although the conjunction of 'p v q' and '–p' entails 'q'—because '(p v q).–p.–q' is inconsistent—nevertheless the claim that 'p v q' entails 'If –p then q' is false. The latter would be true if

'p v q' and '–(If –p then q)'

were inconsistent and this would not be the case unless each of 'p' and '–(If –p then q)' and 'q' and '–(If –p then q) were inconsistent. But the confirmation of each of 'p' and 'q' proves it without proving 'If –p then q,' which is unprovable and neither in fact even confirms the conditional in the ways conditionals are said to be confirmable in sections 2 and 4 of this chapter.

The theory that results from this approach to entailment and confirmation preserves for the conditional the validity of contraposition and transitivity and it rejects the inferences from 'p.q' or '–p v q' to 'If p then q'. And in doing this it justifies a common impression of the logical relations of conditionals. OC is thus in conflict with the dominant theory of conditionals that originated with R. C. Stalnaker and that I have described as the 'possible worlds' theory.

On Stalnaker's view, a conditional is to be evaluated in this way: 'consider a possible world in which A is true, and that otherwise differs minimally from the actual world. 'If A, then B' is true (false) just in case B is true (false) in that possible world'.[15] It would seem, by this account, that 'If I were to drop this pen on the floor, then President Bush would be president of the United States' is true because a world that differs minimally from the actual one is a world in which George Bush is president. This is Stalnaker's position, and also that of the other writers who have advanced the possible worlds theory of conditionality. If B is true in the actual world, any statement of the form (If A then B) is true provided there is no causal connection between the phenomena to which A and B refer which would make not-B true if A is true.[16] This is something established beyond doubt if both A and B are true in the actual world. And if A and B are false in the actual world, then (If A then B) is true, by this theory, only if there is a causal connection between the situations represented by A and B. Thus the making of causal judgments is essential to the application of the theory. Let us say that *A* and *B* are the events or situations the fulfillments of which are expressed by A and B. Then, if A is false and B is true in the actual world, then one must judge that (*A* causes not–*B*) is false before proceeding to affirm (If A then B). And if A and B are both false in the actual world, then (*A* causes *B*) must be true if one is safely to populate possible but unactual worlds with As and Bs in such a way as to make (If A then B) true.

But then the possible worlds theory leaves out an important step. Because the truth of (A causes B) licences the implication of (If A then B) in a sense of that conditional that requires that the presence of *A* ensures the presence of *B*. And so we must, of necessity, judge that (If A then B) is true, in a basic and widespread usage of that expression, as part of the process by which we evaluate the possible worlds conditional with the help of Stalnaker's thought experiment. And the assumed conditional is different from the possible worlds conditional because it is incompatible with there being no connection between *A* and *B*, whereas the possible worlds conditional is inclusive of such arrangements. But if this is the procedure that must be followed, the thought experiment is incomplete because the very application of the method involves a judgment as to the truth or falsity of a very obvious form of conditional, the logic of which is not given. The possible worlds theorists thus produce a theory of conditionals that does not explain (and does not recognize) the logic of a familiar type of conditional that is implied by the causal langauge on which the possible worlds theory depends.

And, having once made this leap, the possible worlds theorists are prepared to assert that the features of the conditional they have constructed are really those of the conditionals found in natural language and that we are wrong if we think that natural language conditionals may have logical relations that are rationally acceptable but different from the conditional of their devising. The peculiarities of the conditional that emerges from the possible worlds analysis are claimed to be characteristic of all proper conditionals and are resolutely defended. But if the theory is right, it seems strange to me that these peculiarities should have to be supported by such very odd arguments. I have commented on Stalnaker's criticism of contraposition (pp. 32–37) and on Lewis's defense of the possible worlds claim that 'p and q' entails 'If p then q' (pp. 31–32). We will come upon other arguments against contraposition (pp. 137–8) and against transitivity for conditionals (pp. 72–74) which, if taken seriously, would reduce even the propositional calculus to a state of confusion.

The confirmation–based logic of OC makes it possible to justify our intuitions about the logic of the conditional that is 'ordinary' in the sense I have described. It rejects the entailment of 'If p then q' by 'p.q' or '–p v q' or '–p' but affirms the validity of contraposition and transitivity for conditionals. To my knowledge it is the only theory that does this. It embraces the logic of conditionals implied by causal relations and also patterns that are not obviously causal in import, such as conditionals about human purposes ('If he left a message on the answering machine, then he wants to hear from you') and evidential conditionals ('If the lights are on, then someone is at home'). The OC theory also permits the development of a logic of general conditionals that has linkages with, and makes coherent, the traditional logic of the syllogism.

But the logic of factual conditionals is influenced by considerations of context and various nonformal linguistic rules. In chapter 4 I examine the logic of one kind of prescriptive rule, the convention—to be considered as a type of conditional—and indicate how account can be taken of the influence of rules and of context on the structure of evidential statements, and hence on the logical relations of parent statements. We see in chapter 5 how awareness of conventional and also of conceptually necessary associations helps us to resolve certain puzzles in the theory of confirmation.

The theory of deduction might be described as the bringing into view of those practices or laws that enable us to avoid contradiction in our discourse. This essay is largely a discussion of some of the laws that make possible the coherent use of those nonelementary expressions we call conditionals. The final chapter is an examination of the logic of those laws themselves, of the logical relations between logical statements. In this logic, entailment (formal deduction) appears as the same relation that controls deductive linkages between factual statements and between conventions. And conditionality is revealed to be an important tool in the expression of logical theses. The course of the argument points to the conclusion that the logical conditional and entailment have distinct and different functions.

Chapter 2

SINGULAR CONDITIONALS

I

The sentential letters 'p', 'q', 'r', . . . will be called the 'elements' of a formulation. A 'component' of a formulation is an expression that may be interpreted by a statement. An element of a nonconditional formulation may be interpreted by a statement and is therefore a component. A singular conditional is a construction out of sentential elements. It consists of an assignment of a sentential element or group of elements to its condition (condition clause) and a similar assignment to its conclusion (conclusion clause), and it is represented by the form '. . .[———]' in which the part outside the square bracket is the condition and the part inside the bracket the conclusion. An interpretation of the factual conditional form 'p[q]' asserts one thing, that the fulfillment of 'p' would (nondeductively) ensure the fulfillment of 'q'. It does not assert either the truth or falsity of either 'p' or 'q'; the forms, 'p' and 'q', and their denials, when embedded in a conditional, are elements of the conditional but are not components. They do, however, become interpretable by statements when they become parts of nonconditional evidential expressions asserting the presence or absence of confirming or disconfirming situations. A conditional, on the other hand, is itself a component. Elementary nonconditional sentence forms such as 'p', 'q', . . . and conditionals such as 'p[q]' and 'pq[r v s]' are the components of the compounds in which they occur. The word 'element' will be reserved for sentential elements of the type 'p', 'q', and their negations, of which components are made up. Thus the sentential element 'p' is a component when it represents an elementary statement, but is part only of the single component 'p[q]'. I have noted that the constants '-', 'v', and '.' are to receive the treatment they customarily receive in bivalent systems.[1] Elements placed side by side, for instance, 'pq', will be read as conjoined. Again, italicized elements, *p*, *q*, . . . represent the situations to which 'p' and 'q' refer.

I have chosen the distinctive notation '. . .[———]' to represent the conditional form of OC, instead of the arrow, ———>, or one of the other signs used in studies of the conditional, because OC distinguishes a conditional that has not previously received separate examination and because the OC approach to conditionality is sharply different from those that have been taken by other writers. I wish to forestall the inadvertent adoption by the reader of misleading assumptions about the logic of the OC conditional that might be prompted by a more familiar symbolism.

In approaching the logic of the conditional I will assume, unless the contrary is stipulated, that what I have called the 'relevent context', the part of the conditional's background assumed to be relevant, does not preclude the occurrence of any combination of condition and result. Thus it is to be accepted that, unless otherwise specified, where the logic of 'p[q]' is in question, none of 'pq', '–pq', 'p–q', or '–p–q' will make up part of the relevant context within the limitations of which evidential support is to be sought: none of them will be excluded from the range of possible confirmation or disconfirmation by a presumption that certain features of the context are 'given.' We want to see the effect the logically possible forms of support and infirmation of 'p[q]' would have on other statements, conditional and nonconditional. In this regard it is helpful, when thinking of the type of support a conditional of the form 'p[q]' may take, to consider the conditional as being in the present indicative and as having reference to a single occasion, one phase of which is in the present and the other of which is in the future. The occasion as a whole is to be conceptually open to occupation by any combination of the presences or absences of p and q. The application of the conditional is not to be thought of as restricted, as in the case of the Gordon Liddy example in section 2 of chapter 1, by virtue of the meanings of the expressions it contains. For the purposes of representing the logic of the conditional, the occasion to which it has reference is taken as divisible into as many segments or phases as may be required to accommodate the confirmation of the conditional in each alternative way. The logic of 'If the block is removed from the wheel, the wagon will roll down the hill' can only be revealed if the occasion of which it is asserted can be divided into two segments, during one of which we can see that when the wagon is not moving the block has not been removed, and during the other of which the block has been removed and the wagon is rolling. The possibility of confirmation in each alternative way is not always fulfilled for singular conditionals. In what follows, however, 'the logic of the conditional' will import, unless the contrary is indicated, the logical relations of the conditional in optimal environments.

In discourse we usually indicate contextual limitations by our manner of giving expression to the conditional relation. If we know that 'p' is true, the conditional relation between *p* and *q* is expressed by 'p is true, so q is true' or 'p hence q'. If, however, the context includes the absence of *p*, we express conditionality by the counterfactual, 'If p had been so, q would have been so'. But the counterfactual should not be confused with the ordinary subjunctive, 'If p were the case, q would be the case' which, like the indicative conditional asserted in ordinary circumstances, assumes a context that does not either preclude or specify from the outset the occurrence or absence of either *p* or *q*. In fact we can often test the appropriateness of the assertion 'If p then q' as an indication of our view of a situation by asking ourselves if it would be correct to assert 'If p were the case, then q would be the case'. I shall argue in chapter 4 that objections to the treatment of 'If p then q' and 'If p were so then *q* would be so' as equivalent rely on differences of relevant context. It is not the precise visible form of the conditional that matters so much as what one says with it. The assertion that if *p* is fulfilled, this ensures that *q* is or will be fulfilled may be expressed as well by the indicative as by the subjunctive form.

Recognition of the fact that there is not one form of conditional only but probably several, each with distinctive features, has become widespread. No attempt is made in this essay to capture in a formal way the peculiarities of conditionals other than those belonging to one or other of the varieties of the ordinary conditional. In this and the succeeding chapter I devote attention to the factual conditional, in its singular and general forms, of which it is true to say that fulfillment of the condition warrants the fulfillment of the conclusion but that remains problematical notwithstanding confirmation. We should, however, be aware of some of the differences between what I have called 'ordinary' and certain other familiar forms of conditional so that we will not find ourselves trying to illustrate the logic of ordinary conditionals with examples that belong to different types.

It can be said of some conditionals that they are obviously not of the kind I have described as 'ordinary.' 'If you want my opinion, he's a mere dilettante' and 'If he only knew, she loves him' belong to a class of conditional of which clearly it would not be correct to say that the assertion that the conclusion is true is in any way justified by the satisfaction of the condition. It also seems conspicuous that the 'even if' conditional is not an ordinary one. With the conditional, 'Even if you wear a coat, you will be cold' one is not saying that wearing a coat will make you cold but that wearing a coat is not sufficient to prevent you from being cold. And what 'Even if you fix the ignition, the car will not start' says is not that fixing the ignition will preclude the starting of the car but that fixing the ignition in itself is insufficient to make it possible to start the car. One indication of the difference between 'even if' and ordinary conditionals is that *modus tollens* does not apply to the former.[2] If you are not cold, it does not follow that you haven't put on a coat; and if the car starts, we cannot conclude that the ignition has not been fixed.

But how is this so? If I maintain that even if you wear a coat you will be cold and also accept that, as it turns out, you are not cold, is not the conclusion that you have not worn a coat the only one that saves the premise? One reason why we do not allow this result may be that if you are not cold, the 'even if' conditional is past saving in any case. And this is because the 'even if' conditional, for instance, 'Even if you wear a coat you will be cold' implies the fulfillment of the conclusion—that you will be cold. Thus, if it happens that you are not cold at the indicated time, the falsity of 'Even if you wear a coat you will be cold' cannot be averted by your not wearing a coat; because, while implying that wearing a coat will not be sufficient to prevent you from being cold, the statement also asserts with some definiteness that you are going to be cold.[3] The antecedent, 'Even if you wear a coat you will be cold' becomes, when 'You do not become cold' is added to it, a contradiction, and as such entails, in standard logic, any consequent, including both 'You wear a coat' and 'You do not wear a coat'. A contradiction will not support a sound deduction.

Another possible view of the logic of 'Even if p then q' is that it is equivalent to the denial of 'If p then –q'—that it denies that if *p* occurs then *q* will not occur. This version would also explain why *modus tollens* is not a feature of the 'even if' conditional. From 'It is false that if you wear a coat, you will not be cold' and 'You do not become cold', one cannot deduce 'You do not wear a coat'.

These thoughts, I admit, do not give us a very satisfactory idea of what the logic of the 'even if' conditional is like. But they are sufficient to show that this logic is different from that of the ordinary conditional and that examples of 'even if' conditionals cannot be counted on to serve as illustrations of the logic of ordinary conditionals. There is another very common form of conditional that differs from the ordinary one in more subtle ways. This is the form I call the 'disposition' conditional, the one we use to define what are usually termed 'disposition predicates'. We explain the meaning of 'This is buoyant in water' by 'If this is in water, it floats'; of 'This is soluble in water' by 'If this is in water, it dissolves'; and of 'This is elastic' by 'If this is pulled, it stretches'. I am speaking here of a limited type of disposition conditional—one that expresses a disposition that is fully revealed by a test and result. It belongs to a class that does not include, for instance, those conditionals that express what are sometimes called 'behavioral dispositions' and that are discussed in chapter 4. The latter are usually genuine ordinary conditionals and are different from the 'defining' type of disposition conditional for a reason I shall explain. The defining disposition conditional is not distinguishable in appearance from the ordinary one, yet it has quite different logical properties.

For one thing, to say that an object is not buoyant in water means that if you place it in water it will not float. Therefore, if, 'If you place this in water, it will float' is to serve as an explanation of the meaning of the disposition term 'buoyant in water', the denial of the conditional must entail 'If you place this in water, it will not float'. But we have seen, with regard to the ordinary conditional, that the denial of 'If p then q' does not entail 'If p then –q'. Thus if one wishes to deny that Green is influenced by Brown to such an extent that if Brown buys a condo, then Green will, he is not bound to say that if Brown buys a condo then Green will not. Green may buy a condo without regard to what Brown does. A second difference between ordinary and what I have called 'disposition' conditionals is that although one can say of the ordinary conditional 'If p then q' that it may be false notwithstanding its complete confirmation, this is not the case for the corresponding disposition conditional. If a thing not in water is placed in water and found to float, this proves it is buoyant in water—we cannot accept the truth of the confirmation and reject the disposition conditional it supports. In this respect it is like the 'betting' conditional I mentioned in the Preface.[4]

The distinction between the disposition and the ordinary conditional can be partly explained in terms of a difference in their respective functions. A disposition term expresses a property capable of being observed under testing circumstances that may or may not be fulfilled, and the disposition conditional indicates the circumstance which, if fulfilled, yields the property to be observed. The circumstance is itself part of the complex observable event that constitutes the disposition characteristic. When the conclusion is fulfilled on the occurrence of the conditioning event, everything the disposition conditional has said is open to observation. When a thing floats in water, the question of whether its being in water justifies the conclusion that it floats in water does not arise because this is not what a disposition conditional is designed to tell us. The dispositional assertion indicates what circumstance leads to the manifestation of a complex characteristic that includes the circumstance itself. That this circumstance is part of the dispositional characteristic is often made explicit in the expression of the conclusion: a thing that floats in water is in water, as is a thing that dissolves in water. In this respect the disposition conditional is like the 'suppositive' conditional (see p. xiv) in that the predicates of the condition and conclusion clauses are so related that the fulfillment of the condition clause is, for semantical reasons, essential to the fulfillment of the conclusion clause, as in, for instance, 'If it snows, it will snow heavily'. The status of the suppositive conditional will be discussed in chapter 4 (see pp. 137–8).

Once the complex characteristic addressed by the disposition conditional has been instantiated, the disposition conditional is conclusively established, not subject to amendment in the light of future experience. But it is an object of ordinary conditionals to give expression to features of the design of things, and judgments about design are always open to revision. Thus conditionals that assert behavioral dispositions are like ordinary ones in that they are not proved by confirmation and are subject to reassessment with regard to future considerations. Notwithstanding that 'If Jane flirts with Paul, John will become angry' is confirmed by Jane's flirtatiousness and John's anger, though John was not angry before flirtation commenced, it may still be arguable that John's anger resulted from features of the situation not involving Jane's behavior.

It is because one's assessment of the truth of a disposition conditional, once it has been confirmed, is not subject to correction that its confirmation may be accepted as conclusive. In chapter 4 I will suggest a reason why the denial of a disposition conditional entails its opposite—why, where 'p[q]' is a disposition conditional, '–(p[q])' entails 'p[–q]' (see pp. 136–7).

'Even if' and disposition conditionals have things to say dissimilar from what is said by the ordinary conditional and they differ in their relations to the events that directly support or undermine them. And although no attempt is made in this essay to formalize the logical relations of conditionals other than the ordinary one, it is important to bear in mind that there are several types of conditionals in common use. Because if we ignore the differences between types of conditionals and try to illustrate the logical features of ordinary conditionals with examples that have unlike relations to the circumstances of their confirmation and disconfirmation, only muddles will result.

I have taken, as examples of ordinary conditionals, assertions about physical nature and also about social relations. Sometimes it has been appropriate to say of a factual conditional of the form 'If p then q' that the fulfillment of the condition would 'bring about' or 'give rise to' the fulfillment of the conclusion; of other examples it would be more correct to say that *p* is a 'symptom' of *q* or that the presence of *p evidences* the presence of *q*. Treatment of all of these examples as ordinary factual conditionals is acceptable if in each case what is being said is that the fulfillment of the condition ensures the fulfillment of the conclusion and if it is allowed that, although disconfirmation disproves the assertion, even complete confirmation does not prove it. The ordinary conditional as it were stands as a conjecture the truth of which cannot be deduced from the direct evidence in its support.

If one accepts the principle of the uniformity of nature, it would be tempting here to argue that if the singular conditional 'If p then q' is true this must be because the occurrence of *p* always justifies the conclusion that *q* will occur or has occurred and that therefore what would entitle us to to deny the truth of 'If p then q' notwithstanding its complete confirmation would be the disconfirmation of some universalization of 'If p then q' that stands behind it. And it is true that we tend to discount the likelihood of the truth of singular conditionals about physical nature, even when confirmed, if universalization of them can be shown to have been disconfirmed. This is what we often mean when we say that a singular conditional, although completely confirmed, has been made questionable by other evidence. In such cases we may say that the fulfillment of 'q,' given 'p,' is at best only probable. But it is far from clear that all singular conditionals are susceptible of this type of analysis. Conditionals about human behavior, about what a person will do given certain conditions, may be accepted as true although their universalizations are false. This would be the case if they are based on assessments of the person's intentions or purposes. (A) 'If Smith's net taxable income is $20,000, this is the amount of income that will be declared in his tax return' may be true if it is Smith's intention to make a true declaration to the tax department. But it may not be true that 'Whenever Smith's net taxable income is $20,000, this is the amount that will be declared in his tax return' because Smith's intentions may change. Some would argue that the universalization of (A) will be made true if Smith's intention is included in the condition; yet others will reply that if the condition includes the purposive factor itself, the assertion as a whole becomes analytic, conceptually necessary, and loses its status as a factual claim. The issue is a vexed one and is not perhaps capable of being resolved to everyone's satisfaction. It thus seems best, for the purposes of logical theory, to permit the singular factual conditional to stand on its own, and to allow that it does not formally entail its own universalization, although the universal conditional (as we shall see in chapter 3) does entail its singular applications.

Before proceeding, we should remind ourselves that 'If p then q' does not always involve a conditioning event that is to occur before the indicated result.[5] The 'then' of the conditional does not imply a forward direction in time. If a given conditional does involve a temporal element, that element attaches to the particular conditioning event and projected event and travels with them. Thus the contrapositive of (1) 'If I open the sluice then water will flow from the dam' is not 'If water does not flow from the dam then I will not open the sluice,' but (2) 'If water does not flow from the dam then I have not opened the sluice'. (1) and (2) might be as well expressed by (1)(i) 'If I open the sluice (t–1) then water flows from the dam (t–2)' and (2)(i) 'If water does not flow from the dam (t–2) then I do not open the sluice (t–1).'

II

We saw in chap. 1 that a factual statement, to *be* factual, must be so related to observables that evidential statements expressing the presence or absence of these observables will justify or lend support to the affirmation or denial of the statement asserted. Factual statements are 'risk laden' in this sense. Nonconditionals differ from conditionals in that the risk to which a nonconditional is subject is certain to be realized. When the occasion to which the nonconditional has reference comes to pass, some conjunction of the elements or their negations contained in it will occur, and this conjunction will prove or disprove it. If 'p v q' is asserted, for example, the occasion to which it refers will exemplify '–pq' or 'pq' or 'p–q,' in which case 'p v q' is proved, or it will exemplify '–p–q,' in which case 'p v q' is falsified. But not every possible combination of elements confirms or disconfirms the conditional. For instance, if '–pq' is exemplified throughout the occasion, 'p[q]' will be neither confirmed nor discomfirmed. We will say that it is only when a phase of an occasion fulfills a 'risk factor', namely, all but one of the elements in a conjunction that will disconfirm the conditional, that the conditional is 'tested', exposed to the risk of falsification. If the denial of the remaining element is fulfilled, the conditional is partially confirmed; otherwise it is disconfirmed and disproved. The disconfirmation of 'p[q]' has the form 'p–q'. Taking (i), (ii) . . . again as indicating different phases of an occasion, 'p(i)' constitutes all but one of the elements in 'p–q(i)' and is therefore one of its risk factors. If, given 'p(i)', 'q(i)' is true, the risk is discharged and the conditional is partially confirmed by 'pq(i)'. The other risk factor '–q(ii)' introduces a risk that is discharged by the fulfillment of the remaining partial confirmation form '–p–q(ii)'.

The disconfirmation of a conditional is formed by conjoining the condition with the denial of the result. Thus, 'pq[r]' is disconfirmed by an occurrence of 'pq–r' and its risk factors are 'pq(i)', 'q–r(ii)', and 'p–r(iii)'. The jeopardy created by the fulfillment of these risk factors is discharged, respectively, by 'pqr(i)', '–pq–r(ii)', and 'p–q–r(iii)' during appropriate segments of the occasion in question. Any disconfirmation may be reduced to a disjunction of conjunctions by

(a) applying de Morgan's laws and the law of double negation to remove all negations outside brackets and
(b) using the distributive law to produce the required form.

Therefore, the disconfirmation form of 'p[qr]' is '(p)–(qr)'. One application of de Morgan's law as to the negation of conjunctions yields '(p)(–q v –r)'; and one application of the distributive law gives us 'p–q v p–r'. The fulfillment of either conjunction disconfirms 'p[qr]'. Thus a conditional may have more than one form of disconfirmation.

Nonconditionals report that things have certain characteristics; conditionals assert relations between things having such characteristics. Singular nonconditionals, and their denials, are reducible to conjunctions, or disjunctions of conjunctions, of elementary statements, and these formulations are the proofs and disproofs of the nonconditionals. Conditionals are confirmed by phased conjunctions of elementary statements, but these conjunctions, being about occurrences and not about relations between occurrences, are not reductions of the conditionals they confirm and do not prove them. If logical relations between statements are to be determined by reference to their confirmations and disconfirmations, then if we are to establish the logical relations between conditionals and nonconditionals, we must somehow show the combined effect of confirmations and disconfirmations of nonconditionals during occasions and of conditionals during phases of occasions. How do we do this? I treat a nonconditional elementary expression, for instance, 'p', as holding throughout the occasion to which the entire formulation of which it is a part applies, and a conditional form as holding throughout this occasion as well. When a conditional is conjoined with a nonconditional component or components its confirmations are circumscribed by the fact that the nonconditional components of the conjunction are taken as true throughout the occasion; the conditional may then be confirmed in a way or ways not inconsistent with the truth of the nonconditional component or components. Thus, in 'p.p[q]', confirmation of 'p[q]' is confined to the form 'p . . .q . . .' which must, like 'p', hold during the whole occasion. Confirmation of the conditional during a phase 'n' of the occasion by '–p . . . –q . . .' would yield the contradictory form 'p–p–q(n)'. Conditionals themselves, if true, are true of the occasion as a whole although they are ideally confirmed by (but not reducible to) statements about features of parts of the occasion. Singular statements that enter into entailment relations with each other are assumed to hold, as we have seen, during the occasion to which they refer. Because phased confirmations, for instance, 'pq(i)', '–p–q(ii)', express situations that exist during parts only of the occasion, they may not participate directly in entailments, notwithstanding that they are essential to the process by which entailments are determined.

III

One of the consequences of the 'riskable' nature of factual statements is this. If an assertion is to qualify as factual, it must be linked to events the occurrence of which would confirm it and to other events the occurrence of which would disconfirm it. An unconditional form must be interpretable by assertions that are exposed to some risk; schemata such as 'p.–p' and '–p v p,' which are not so interpetable, which must take either contradictory or tautologous confirmations or disconfirmations, cannot stand as expressions having entailment relations with other expressions. They fail what I have referred to in chapter 1 as the 'application requirement.' Conditional statements, we have seen, need not be exposed to risk by the circumstances in which they are asserted; but they must be vulnerable to confirmation or disconfirmation in the event that their risk factors are fulfilled. And this cannot happen if the confirming or disconfirming result follows logically from the fulfillment of the risk factor. If this is the case then we say that the conditional is not well formed.

Thus 'p[p]', with risk factors 'p' and '–p' and disconfirmation 'p.–p', cannot be the form of a factual conditional because if it is placed in jeopardy by either 'p' or '–p,' the denial of the remainder of its disconfirmation is truth functionally implied. A conditional construction only one of whose confirmation forms is defective is also not well formed.[6] The proper form of the truth sought to be expressed in 'p[p]' is 'p entails p', which is an unconditional, nonfactual, and also nontautologous statement of logic.

It will seem odd to deny that 'p entails p' is a tautology. But it is not a tautology if we understand 'A is a tautology' to mean simply '–A is a contradiction'. What 'p entails p' says is that 'p' is deducible from 'p' pursuant to some theory of deduction one is prepared to accept. And although it would be absurd to deny this entailment, it would not be strictly contradictory to do so; 'p entails p' would be put in doubt if we could advance an interpretation that shows us how we might intelligibly both accept and reject the assertion of 'p'. But this would be an argument about logic. In using 'p entails p' we do not suppose, as we would in the case of a factual conditional, that it would be rational to seek out disconfirmation by bringing about *p* and watching to see if *p* is absent. The question is the logical one of whether, if 'p entails p' were abandoned, our speech would still have intelligible consequences. I discuss at greater length the nature of entailment statements in chap. 6.

No conditional is well formed if it contains elements that make either its disconfirmation or its confirmation inconsistent. Another form of this type is '–p[p]'. Here the problem is that the risk factor truth functionally (TF) implies the remainder of the disconfirmation instead of the denial of the remainder, and so each of its confirmations is self-contradictory. Aristotle held[7] that 'If not p then p' is an absurdity and has been frequently criticized for so doing.[8] And *if* Aristotle, consciously or otherwise, meant by 'If not p then p' simply 'p v p', then his critics are right, because 'p v p' is merely a redundant way of saying 'p', which is not an absurdity but clearly eligible for interpretation by true or false statements. It seems much more likely, however, that Aristotle took the simple conditional form to say that the fulfillment of the condition is sufficient to ensure the truth of the conclusion. And if this is the case, he was right to say that 'If not p then p' is absurd.

IV

The denial of a conditional is confirmed by what would disconfirm the conditional and disconfirmed by what would confirm it; '–(p[q])' is confirmed by the fulfillment of the form 'p . . . –q . . .' and disconfirmed by the fulfillment of the segmented forms 'pq(i)' and '–p–q(ii)'. But although a conditional entails the denial of its own disconfirmation, the negation of the conditional does not entail its own confirmation. 'If you do a dance, it will rain' entails that it is not the case that you have danced and it has not rained. But it does not follow from 'It is false that if you do a dance it will rain' that you have danced and it has not rained. The denial of a conditional does not necessitate the occurrence of any particular event; rather it denies the existence of a natural connection between the elements in the condition and the conclusion. The form, '–(p[q])' means that if *p* occurs, one cannot on that account say that *q* will occur. It does not mean that the form 'p . . . –q . . .' has been fulfilled. Nor does '–(p[–q])' mean that if *p* occurs, *q* may or might occur, in the sense of 'may' or 'might,' which implies that the occurrence of *p* is relevant to, or facilitates, the occurrence of *q*. 'If you let the engine dry out (p), it might start (q)' may be used as a primitive form of probability statement that implies that the occurrence of *q* is made more likely by the occurrence of *p*. But this is not the import of '–(p[–q])' in the present essay. The translation of '–(p[–q])' by 'If p is true, q might be true' is acceptable only if 'might' is used merely to signify not the probability that *q* will occur, given *p*, but the absence of a natural connection between the occurrence of *p* and the occurrence of –*q*. The forms, '–(p[–q])' and '–(p[q])' deny relations; they do not assert new ones.

The confirmation of the denial of a conditional proves it. Instantiation of 'p . . . –q . . .' puts the truth of '–(p[q])' beyond deniability. Its disconfirmation, however, does not disprove it. If I do a dance today and it rains tomorrow, these events constitute, together with my not having previously danced and its having remained dry, the direct infirmation of 'It is false and if I do a dance it will rain'; but they do not establish its falsity. The view that my dancing had nothing to do with bringing on the rain remains highly plausible.

A conditional is tested by the occurrence of all but one of the elements in a conjunction that would disconfirm it; the denial of the conditional is tested by the occurrence of all but one of the elements in a conjunction that would confirm it. But if the conditional fails the test, it is disproved, whereas failure of a test by the denial of a conditional does not disprove the denial although it infirms it, makes its truth less likely. Thus, placing a form in a state of risk, putting it in peril of being shown to be false, implies that it is being tested. But to say that a form is tested is not necessarily to say that it has been placed in a state of risk, in the sense of 'risk,' which implies that the statement is disproved if it fails the test.

It is worth observing that where a conditional has but one form of disconfirmation, then because the disconfirmation of the conditional is the same conjunction as the confirmation of its denial, each has the same test factors. And the element which, when added to the test factor confirms the one, is the negation of the element that confirms the other. Thus, 'p[q]' is disconfirmed by the instantiation of 'p . . . –q . . .' and this is the confirmation of '–(p[q])'. The form, 'p[q]' is tested by the presence of all but one of the elements in 'p . . . –q . . .' and so is '–(p[q])', that is, by 'p . . .' and by '–q. . . '. So, while 'p[q]' is then disconfirmed by the form 'p . . .–q . . .'. '–(p[q])' is confirmed by it and while 'p[q]' is confirmed by the phased forms 'p . . . q . . .' and '–p . . .–q . . .', these are the infirmations of '–(p[q])'.

Consider, 'If John is encouraged he will work hard' (S(i)). This is receiving partial confirmation at the moment because he has not been encouraged and is lazy, and will receive part-confirmation in an alternative way if he works hard when encouraged. It will be disconfirmed and disproved if he does not work hard after encouragement. 'It is false that if John is encouraged he will work hard' (S(ii)), is proved if he does not work hard when encouraged and is tested by what tests S(i): by John's not working hard and by his being encouraged. The full infirmation of S(ii) is fulfilled by the events that confirm S(i); but infirmation of S(ii) does not disprove it. If John applies himself after being encouraged, it remains possible that the behavior change was brought about not by the encouragement but by the presence in the situation of some other influence.

V

Assuming all this to be the case, how do we assess the claim that a conditional or group of conditionals or a mixed group of conditionals and nonconditionals entails another statement? Let us call the left side of an entailment the 'antecedent' and the right side the 'consequent'. Then, in accordance with the principles discussed in chapter 1, an antecedent A entails a consequent B if and only if both A and B are contingent and A.–B is inconsistent. A test for entailment will become available, therefore, if and only if we can construct a test for inconsistency.

A preliminary step in determining the inconsistency or otherwise of a formulation will be its reduction to a disjunction of conjunctions by the application of de Morgan's laws and by the law of double negation to eliminate all negation signs in front of brackets and then the distributive law to produce a disjunction of conjunctions. Thus an application of the law of double negation, and of de Morgan's law as to the negation of conjunctions, to

— —(A.B).–(C.D) yields
(A.B)(–C v –D).

And the use of the distributive law then gives us

A.B.–C v A.B.–D.

If each of A.B.–C and A.B.–D is inconsistent, then – –(A.B).–(C.D) is inconsistent. But if A, B, C, and D are not all elementary, we cannot decide the question of the inconsistency of these conjunctions by simple inspection for the occurrence of the form 'p.–p.' (Henceforth quotation marks around formulas such as this are dispensed of whenever the context insures that confusion between use and mention will not arise.) The expression

p.–q.p[q]

is surely inconsistent, although it does not conjoin a component with its own denial.

We have seen that the question of the inconsistency of conjunctions containing conditionals is made complex by the fact that conditionals take more than one form of confirmation and the fact also that while conditionals are capable of disproof they are not provable and although their denials are capable of proof they are not disprovable. And the suggestion I made in chap. 1 was that a conjunction is to be deemed inconsistent if and only if it contains within it a group of components such that

(i) the proof of any provable component in the group disproves the conjunction of the rest of the group and
(ii) each component in the group would be disconfirmed, to at least equal degree, by the systematic confirmation of the rest of the group.

I showed in chapter 1[9] that by this criterion –p does not entail p[q] because –p. –(p[q]) is not inconsistent: –p does not TF imply the confirmation of p[q], nor does the proof of –p disprove –(p[q]), which is not disprovable. The test, however, does establish that p[q].–(–q[–p]) is inconsistent and therefore that p[q] entails –q[–p]. I here set out the proof of this entailment in an arrangement that I shall use hereafter. (1) I first place the proposed entailment on a line. (2) Under that line is set out the disjunction of conjunctions (called 'Co') which must be inconsistent if the entailment is to hold. At this stage one considers whether each conjunction in Co satisfies part (i) of the test. In the case at hand we have the one termed disjunction of conjunctions consisting of the conjunction p[q].–(–q[–p]). Its only provable component is –(–q[–p]) and this is proved by instantiation of –q . . . p . . . that TF implies the disproof of p[q]. The conjunction therefore satisfies requirement (i).

We determine whether the conjunction passes test (ii) by finding out whether each component would be disconfirmed by the confirmation of the rest of the conjunction. Let us call any group of all but one of the components in Co an 'implicans' (designated by 'I') and the negation of the remaining component the 'denial of the remainder' or 'DR'. The conjunction will pass test (ii) if the confirmation of each I confirms DR to at least equal degree. Accordingly, at stage (3) of the procedure we place each implicans opposite its corresponding DR and determine the effect on the DRs of the confirmation of the Is. The implicans is placed on the left, the DR on the right and an arrow '⟶ ' will point from left to right if it is being claimed that when the confirmation of each component of the implicans has been systematically carried out, the results will TF imply each form of confirmation of DR. Under the expressions connected by an arrow we put the demonstration that the confirmation of I TF implies the confirmation of DR. If I can be confirmed in more than one way, we divide the demonstrations into segments or phases (i), (ii), (iii) . . . representing the different segments of the occasion during which the different confirmations may be realized. The successive juxtapositions of Is and DRs will be numbered 'D(I)', 'D(II)'. . . . The whole proof, designated 2(1), with stage numbers at the far left, is as follows:

Stage		
1	2(1)	p[q] entails –q[–p]
2		Co is p[q].–(–q[–p])

We see that Co satisfies the first test on inconsistency because confirmation of its only provable component –(–q[–p]) during any phase by the fulfillment of –q . . . p . . . also proves the denial of p[q]. Confirmation during one phase of –(–q[–p]) is sufficient because it is conclusive and makes occurrences at any other phase irrelevant. We may now ask whether Co passes the second test on inconsistency. Let us place the phase indicator on the left of a line, with the intent that the formulations on the rest of the line are to be fulfilled during that phase.

Stage				
3	D(I)	I		DR
		p[q]	——>	–q[–p]
	Phase (i), confirm p[q] by pq	pq	TF implies	pq
	Phase (ii), confirm p[q] by –p–q	–p–q	TF implies	–p–q
	D(II)	I		DR
		–(–q[–p])	——>	–(p[q])
	Confirm –(–q[–p]) by its only confirmation form –q . . . p . . . during any phase	–qp	TF implies	–qp

I have just noted that Co passes test (i). At stage (3) we see that it also passes test (ii). In D(I) the disconfirmation of p[q] has the form p–q and its phased confirmations are therefore p . . . q . . . and –p . . . –q . . . . The form, –q[–p] has the same disconfirmation and therefore the same confirmations. In D(II) the confirmation of –(–q[–p]) is the disconfirmation of –q[–p], –q . . . p . . ., and it TF implies the confirmation of –(p[q]). Again, because the confirmation of –(–q[–p]) is conclusive, only the phase containing it need be reviewed.

Recall that statements on the entailment line hold throughout the occasion that the formulation as a whole is about, and therefore phased confirmations, such as 'pq(i)', may not appear on that line.

Before proceeding further toward the final statement of the definition of entailment, let us examine more closely the question of inconsistency. Suppose we wish to discover whether a conjunction of m components is inconsistent. We first test all the groups of two components contained in the conjunction; if none of these groups is inconsistent, we test each group of three components and so on. If we find a group of n members (n equal to or less than m) which is inconsistent in this way we have what I shall call a 'normal' inconsistency, namely, one of which it is true to say that each group of n–1 of the n components contained in the n membered conjunction is consistent. Every conjunction that contains a normal inconsistency is itself inconsistent. What are the characteristics of a normally inconsistent conjunction of n members?

(1) The systematic confirmation of each component of a conjunction of n–1 members (each component of an implicans) must result in confirmation of the denial of the remaining component (the DR). By this I mean that if we take each component of the implicans and project its confirmation in each alternative way, at the end of this process the denial of the remainder must have been confirmed in each alternative way. It is not that the confirmation of each component must have this result but that the result will have been achieved at least by the time the process has been applied to each component. And the confirmation of a component can only take effect in ways that do not bring about the disconfirmation of any other *disprovable* component of the implicans. In other words, since one is trying to establish that the confirmation of the implicans makes the truth of the implicans incompatible with the falsity of the denial of the remainder, one must not permit the confirmation of one component to affect the result if such confirmation renders the truth of the implicans impossible through the disproof of one of its other members. Imagine that one wants to establish the effect of the confirmation by p . . . q . . . of p[q] within the implicans

p[q].q[r].r[s].

Each of q[r] and r[s] is disprovable and so the confirmation of p[q] must be accompanied by the denial of the disconfirmation of the other two. The testing of the whole by the confirmation, during one segment, of p[q] by pq(i) then must take the form,

(i) (pq)(–q v r)(–r v s).

The confirmation of a form potentially imperils its disprovable associates and 'forces' the avoidance of disconfirmation. A nondisprovable component cannot be forced in this way because its disconfirmation is not incompatible with its truth. Thus in,

–(p[–q]).q[r]

confirmation of q[r], say, by –q–r(i) does not force the nondisconfirmation of –(p[–q]) because this is not disprovable. The confirmation, and proof, however, of –(p[–q]) by pq(ii) would force the disprovable q[r] so that confirmation of –(p[–q]) bears on the whole implicans to produce

(ii) (pq)(–q v r).

(2) Notice that in (1) the function of confirmation of the members of the implicans is to produce the confirmation of the denial of the remainder. It is not sufficient that it implies simply the negation of the confirmation of the remainder. This requirement is a consequence of the fact that the denial of the disconfirmation of an expression is not always equivalent to its confirmation. Thus –p, although incompatible with the confirmation of –(p[q]), does not TF imply the confirmation of p[q] and so although if –p is true it is not possible for us to prove the denial of p[q], we do not, on the ground alone of –p, have support for the claim that p[q] is true.

(3) In addition, it must not be possible systematically to confirm *any* n–1 membered group, any implicans, without confirming the denial of the nth member of the n-membered conjunction. This requirement is also imposed by the difference that exists between the confirmation of a conditional and the denial of its disconfirmation. And failure to satisfy the prerequisite is again illustrated by the case of –p and p[q]. Disconfirmation of the latter disconfirms the former, but, because the confirmation of p[q] is not equivalent to the denial of its disconfirmation, confirmation of –p does not confirm p[q]. If the requirement is not satisfied, then, it is possible that nature will supply us with a combination of events that will support

n-1 members of the conjunction without supporting the denial of the nth member.

(4) If a DR takes more than one form of confirmation, then the phased or segmented confirmation of the implicans must confirm the DR in each alternative way. If this does not occur, we cannot be sure that confirmation of I will be accompanied by the sort of confirmation of DR that affords it significant support. It is not necessary that each confirmation form of the implicans TF imply a confirmation form of DR, but just that when the implicans has been confirmed in each alternative way, the DR will also be confirmed in each alternative way.

(5) The proof of the implicans must prove the corresponding DR. If we receive proof of the implicans we must not be in a position to doubt the DR.

(6) If any DR is disprovable, its disproof must TF imply the disproof of its remaining members. In other words, the proof of any member, conjoined with the denials of the disproofs of the other members, must yield a contradiction of the basic truth functional type: it must produce a one or more termed disjunction of conjunctions, each conjunctive member of which has the form of A.–A. . . .

The definition of entailment is based upon a definition of inconsistency that meets these conditions; it is a definition that permits us to test for inconsistency conjunctions of more than two components, and I ask the reader to use it hereafter in preference to the less precise account of entailment given in chapter 1.

Entailment. One set of statements (the 'antecedent') entails another (the 'consequent') if they are each contingent and if each conjunction in the disjunction of conjunctions formed from the affirmation of the one set and denial of the other is inconsistent. A conjunction is inconsistent if it contains a normal inconsistency and a normal inconsistency is such that

> *E(i)* the confirmation of each of its provable components TF implies the disproof of the other components, and *E(ii)* the systematic confirmation of each implicans TF implies the confirmation, in each alternative way, of its corresponding DR.

This is not to say, with respect to *E(ii)*, that *each* confirmation form of the implicans must TF imply one of the confirmation forms of the DR, but that by the time the implicans has been confirmed in each alternative way, the DR will have been confirmed in each alternative way—which is to say that when each of the confirmation-parts of the implicans has been fulfilled, each of the confirmation-parts of the DR will have been fulfilled.

The question whether an antecedent or consequent is contingent can be settled by applying the inconsistency test to its affirmation and its denial. If either its affirmation or denial is inconsistent, the expression is not contingent.

The definition of inconsistency satisfies the six requirements. Provision E(i) fulfills requirement (6). Provision E(ii) fulfills conditions (1) to (4). E(i) and (ii) satisfy (5). Because of E(ii), confirmation of I confirms DR; and because of E(i), provability of one or more components assures us that all the other components are disprovable and therefore that the denial of each component is provable. E(i) and E(ii) restate the conditions outlined at pp. 65–66 and on which were based the proof, in 2(1), of the entailment of –q[–p] by p[q].

Forcing. When testing an implicans to see if some or all of its confirmations confirm the DR, the confirmations of any component must be accompanied by the nondisconfirmation of any disprovable component of the implicans. As was observed in (1) at p. 67, what is to be shown is that the confirmation of I results in the confirmation, in each possible way, of DR, and one must not select confirmations of one component that render the confirmation of I impossible through the disproof of one of its other members. We want to know the effect on DR of the confirmation of I, not of circumstances that would imply the falsity of I. Thus if the confirmation of a component A is to take part in the confirmation of an I containing a disprovable component B, any usable confirmation of A 'forces' a neutral structure on the elements composing B. If A is –(p[–q]) and B is q[r], and if –(p[–q]) is confirmed by pq(i), this confirmation can only play a role in the confirmation of I if it is compatible with the nondisproof of q[r]—–qv r(i)—during the phase. This is to say that if A is confirmed, and if B is true, then B will not be disproved, and so the circumstances must be compatible with the nondisproof of B—the confirmation of A 'forces' the nondisproof of B.

The confirmation of q[r], however, does not force the nondisconfirmation of –(p[–q]) because the latter is not disprovable. As we have noticed, the denial of a conditional may be true even though such denial has been disconfirmed.

The forcing operation makes interesting entailments possible by bringing into play elements not contained within a single component. Theorem 2(2), showing that conditionality is transitive, illustrates the application of the definition of entailment and also the forcing mechanism.

<u>2(2)</u> p[q].q[r] entails p[r]
Co is p[q].q[r].–(p[r])

> The only provable component is –(p[r]) and each of the others is disprovable. The proof of –(p[r]) and the nondisproof of p[q].q[r] during a phase is a contradiction: (–p v q) (–q v r) (p–r) yields p–r–q–p v p–r–qq v p–rr–p v p–rrq, each conjunctive member of which is contradictory.

		I		DR
D(1)		p[q].q[r]	⟶	p[r]
	Phase (i), confirm p[q] by pq	(pq)(–q v r)	TF implies	pr
	Phase (ii), confirm q[r] by –q–r	(–p v q)(–q–r)	TF implies	–p–r
D(II)		I		DR
		p[q].–(p[r])	⟶	–(q[r])
	Confirm –(p[r]) during a phase by p–r	(–p v q)(p–r)	TF implies	q–r
D(III)		I		DR
		q[r].–(p[r])	⟶	–(p[q])
	Confirm –(p[r]) during a phase by p–r	(–q v r)(p–r)	TF implies	p–q

In D(1), confirmation of p[q] by pq(i) forces the confirmation of p[r] by pr(i). Further confirmation of p[q] by –p . . . –q . . . would force nothing interesting with respect to p[r] and the same holds for confirmation of q[r] by the form q . . . r . . . . The form (–p–q)(–q v r) does not TF imply either of the forms pr or –p–r, nor does (–p v q)(qr). But confirmation of q[r] by –q–r(ii) forces the confirmation of p[r] by

–p–r(ii). And so the systematic, phased confirmation of each of p[q] and q[r] yields the confirmation in each alternative way of p[r]. In D(II) and D(III), the effect of the proof of –(p[r]) on each of its respective conjuncts is to deliver the proofs of –(q[r]) and –(p[q]).

The following example may help the reader see the justice of the result. How do we establish the claim that the conjunction of 'If the fuse burns to its end, the fire will reach the dynamite' (p[q]) and 'If the fire reaches the dynamite, the dynamite will explode' (q[r]) entails 'If the fuse burns to its end, the dynamite will explode' (p[r])? If, surveying the scene before the dynamite has exploded, we see that no fire has reached the dynamite, q[r] is confirmed by –q–r(i). If p[q] is not false, then either the fuse has not burned to its end or it has reached the dynamite. Because the fire has not reached the dynamite, we are compelled to conclude that –p–q–r(i), which TF implies –p–r(i). In another segment of the occasion, we may see that the fuse has burned to its end and has reached the dynamite. If q[r] is not disproved by these events, the dynamite will explode and p[r] will be confirmed by pr(ii). Consider, however, the possibility that the fuse burns to its end and the dynamite fails to explode. Then p[r] is disproved. And if accompanying events do not disprove p[q], –(q[r]) will be proved; and if q[r] is not disproved, –(p[q]) will be proved. Thus the disproof of p[r] disproves the conjunction p[q].q[r].

The conditional is nontransitive in the systems of both Stalnaker and Lewis. Suppose, Stalnaker would say, that 'If p then q' and 'If q then r' are both true because in the closest world in which p is true, q is true and in the closest world in which q is true, r is true. But if the closest world in which p is true is not the closest world in which q is true, then in the closest p world, r may be false. (Lewis's argument would be the same, except that Lewis would refer not to the closest world in which a statement A holds but to some world in which A holds such that B holds in that world and also in every A world at least as similar to the actual world. This difference does not affect the argument against the transitivity of conditionals.) The Stalnaker/Lewis view does not take into consideration the fact that we cannot gain direct evidence of the truth of a conditional except through the occurrences of events in the actual world. One does not collect evidence for the truth of a conditional from occurrences of events in worlds that are in principle nonactual and beyond observation. Actual world evidence for the truth of 'If p then q' that would be countenanced by Stalnaker and Lewis is of the form 'pq', and if this is accompanied by 'r', we have, in Stalnaker/Lewis theory, proof of both 'If q then r' and if 'p then q', and thus also of 'If p then r'. If, on the other hand, the actual world delivers us the disproof of 'If p then r' in the form 'p–r', then if this situation is accompanied by the truth of 'q,' we have the disproof of 'If q then r' and if it is accompanied by '–q,' we have the disproof of 'If p then q'. Thus if we attend exclusively to events occurring or that may occur in the actual world, we cannot, in Stalnaker/Lewis terms, gain direct evidence for the truth of the premises 'If p then q' and if 'q then r' without producing similar evidence for the truth of if 'p then r;' and one cannot gain direct evidence for the falsity of 'If p then r' without producing similar evidence for the falsity of one of the premises. The Stalnaker/Lewis theory, to the extent that its theses are illustrated by actual occurrences, is compatible with the transitivity of conditionals; to the extent that it is inconsistent with transitivity of the conditional, it cannot be illustrated by occurrences capable in principle of being observed.

I do not contend that the Stalnaker/Lewis theory is an implausible or incoherent theory of conditionality. I do suggest that the theory offers a model of conditionals that reflects less well the logic of the conditional of natural language, at least of the type of conditional I have called 'ordinary', than does the conditional of OC. Also, if this conjecture is correct, it is possible that the reason why in ordinary discourse we treat transitivity of conditionals as logically valid is that logical relations in natural language are governed by questions of the occurrences, or possible occurrences, of events in the actual world and not by questions of the occurrences of events in worlds that are in principle beyond human apprehension.

Arguments against the transitivity of conditionals that are designed to put in doubt our acceptance of transitivity in ordinary discourse, standardly use examples that have in common a peculiar characteristic. The following quotation is from Simon Blackburn's *Spreading the Word*(p. 294).

> For example, one might expect this to be a valid form of argument: 'if p then q, if q then r, *so* if p then r'. This is the property of transitivity, and we rely on it quite happily: 'if he comes then she will come; if she comes things will hot up. So if he comes things will hot up.' Encouraged by such examples we might write transitivity into our logic for the connective. But then suppose Smith and Jones are two candidates for an election. We might happily suppose that if Jones wins Smith will be envious

> of Jones. And obviously if Smith dies, Jones will win. But by transitivity the two together yield that if Smith dies, Smith will be extremely envious of Jones.[10]

This example operates, as do many others that can be cited, by making the fulfillment of the condition of one premise incompatible with an assumption that is necessary to the truth of the other premise. Let us treat 'If Jones wins, Smith will be envious of Jones' as 'If q then r' and 'If Smith dies, Jones will win' as 'If p then q'. Clearly, in asserting 'If q then r' we assume that at the time of the fulfillment of q, if it is fulfilled, Smith is alive (s), whereas if the condition of 'If p then q' is to be fulfilled, Smith will at that time be dead. But if we make this assumption that supports 'If q then r' explicit by including it in the condition we get 'If q and s then r' and the principle of transitivity does not license the deduction of 'If p then r' from the premises 'If p then q' and 'If q and s then r.'

I argued in chapter 1 (pp. 41–42) that it is difficult to imagine how any abstract system of logic can be expected to survive a type of analysis that ignores the assumptions embedded in the context in which particular statements are asserted. If, for example, p, q, and r represent these expressions,

(p)(–p v q)(–q v r) entails r

is a valid and unexceptionable deduction in truth functional propositional logic, yet it also produces an absurd result when contextual assumptions are similarly ignored.

VI

The confirmation of unconditional forms will be assumed to prevail during the entire occasion to which they refer. And although individual conditional expressions must be 'perfect' in that all their confirmation and disconfirmation expressions must be contingent, an implicans is accepted as legitimate provided its components are consistent—that is, if they do not entail the denials of each other. As a result, because it is also assumed that all expressions in a singular formulation relate to the same occasion, or period of time, no matter how stretched out, it is sometimes not possible to confirm the conditional in each alternative way consistently with the confirmation of its neighbors in an implicans. We see this effect in entailment 2(3), which demonstrates that *modus ponens* is valid in OC.

2(3) p.p[q] entails q
Co is p.p[q].–q

The conjunction passes the first test on inconsistency. p and –q are provable and p[q] is disprovable. The conjunction (p)(–p v q)(–q) is a contradiction.

D(I)	I		DR
	p.p[q]	⟶	q
Confirm p throughout the occasion	(p)(–p v q)	TF implies	q

p[q] is not confirmable by –p–q because this form of confirmation produces the contradiction (p)(–p–q), which is not a possible form of confirmation. But confirmation of p alone is sufficient to trigger (the single form of) confirmation of the DR.

D(II)	I		DR
	p.–q	⟶	–(p[q])
Confirm p and –q throughout the occasion.	p.–q	TF implies.	–q

D(III)	I		DR
	p[q].–q	⟶	–p
Confirm –q throughout the occasion	(–p v q)(–q)	TF implies	–p

We may use the inconsistent conjunction p.p[q].–q to demonstrate not only *modus ponens*, but also that p[q] entails –p v q, that p.–q entails –(p[q]) and that *modus tollens* holds in OC. All these entailments result from applications of the same inconsistency.

2(4) p[q] entails –p v q (i.e.,–(p–q),
2(5) p–q entails –(p[q])
2(6) p[q].–q entails –p

The conditional of OC bears the logical relations to other conditionals and to nonconditionals that in chapter 1 I suggested it should bear:

1 As is shown by 2(4), the conditional entails the denial of its own disconfirmation.

2 The converse of this is not true: –p v q does not entail p[q]. Neither –p.–(p[q]) nor q.–(p[q]) is inconsistent.

3 Nor does the conditional entail that any of its confirming instances is true. For example, p[q] does not entail pq v –p–q. The disjunction of conjunctions formed from p[q] and –(pq v –p–q) is (p[q]) (–pq) v (p[q])(p–q), and of these, the first is not inconsistent.

4 From the truth of the denial of a conditional, we cannot deduce the truth of its disconfirmation. The form, –(p[q]) does not entail p–q. From the falsity of 'If I jump ten feet in the air, I will turn into a bird' we cannot conclude that, having jumped ten feet in the air, I have failed to become a bird.

5 Also, from the falsity of the conditional we cannot deduce the truth of the opposite claim. The form, –(p[q]) does not entail p[–q]. It is false that if I hit the tree with a stone, Smith will marry Sandra. But this does not permit me to say that my hitting the tree with a stone could prevent the marriage. The stone and the courtship of Smith and Sandra may have, no doubt do have, unrelated histories. As is shown by 2(7), however, p[q] and p[–q] cannot both be true.

2(7)	p[q] entails –(p[–q])
	Co is p[q].p[–q]

Neither component of p[q].p[–q] is provable and so the conjunction fulfills the first requirement on consistency by default.

D(I)	I		DR
	p[q]	⟶	–(p[–q])
Confirm p[q] during a phase by pq	pq	TF implies	pq
D(II)	I		DR
	p[–q]	⟶	–(p[q])
Confirm p[–q] during a phase by p–q	p–q	TF implies	p–q

Each DR is proved by its single form of confirmation. It is not necessary that each confirmation-part of an implicans TF imply the confirmation of its corresponding DR—just that by the time the implicans has been confirmed in each alternative way, the DR will have been confirmed in each alternative way. In 2(7), the DR has in each case but one form of confirmation.

VII

The methods of OC leave unaffected the logic of the propositional calculus to the extent that it applies only to contingent unconditional forms. And the arguments of previous sections have shown that we can establish the truths of,

p[q] entails –q[–p]	[2(1)]	
p[q].q[r] entails p[r]	[2(2)]	and
p.p[q] entails q	[2(3)]	

These statements would be true if the form . . . [———] were replaced by the material conditional form (. . .⊃———). But certain relations that are not entailments under a truth functional treatment of conditionality are entailments in the present system. The formulation (p⊃q)(p⊃–q) is consistent and it is true if p is false. But I have shown in 2(7) that p[q].p[–q] is inconsistent in OC. 2(7) reflects the fact that if we are fortunate enough to collect evidence that justifies the claim 'If ingredient x is administered, the progress of the cancer will abate' we disprove 'If ingredient x is administered the progress of the cancer will not abate'. We have observed also, however, that some entailments that are demonstrable when the conditional is treated as material are not entailments in OC. Although –(p⊃q) entails (p⊃–q), –(p[q]) does not entail p[–q] and although (–p v q) entails (p ⊃ q), it does not entail p[q].

The consequences of the OC treatment of conditionality seem to me to be largely compatible with our ordinary use of conditionals in speech. Three additional examples strengthen this impression.

(1) What is sometimes called the 'compound law of transposition,' that 'If p and q then r' entails 'If p and not r then not q' is provable in the system.

2(8) pq[r] entails and is entailed by p–r[–q]

Each of pq[r] and p–r[–q] has the same disconfirmation, pq–r, and the same alternative confirmation-parts, pqr(i), –pq–r(ii) and p–q–r(iii). Neither is provable and each is disprovable.

(2) The theory sided (see pp. 61–62) with Aristotle in holding that –q[q] is absurd. But the affinities of the theory with Aristotelian propositional logic does not end there. Aristotle had argued that[11] the absurdity of 'If not q then q' has as a consequence that it is impossible that the same thing should be necessitated by the being and not–being of a thing. In our terms, p[q] is inconsistent with –p[q]. Lukasiewicz paraphrases Aristotle's argument in this way. "Two implications of the form 'If p then q' and 'If not–p then q' cannot be together true. For by the law of transposition we get from the first implication the premiss 'If not–q then –p' and this premiss yields together with the second implication the conclusion 'If not–q then q' by the law of hypothetical syllogism."[12] Lukasiewicz rejects the argument because he holds 'If not–q then q' not to be absurd. In the present theory, –q[q] lacks self-consistent confirmation and is therefore not well formed and does not have the opportunity to either entail or be entailed by anything. But it is also true in the theory that p[q] entails –(–p[q]) and so is incompatible with –p[q]. This is shown in 2(9).

2(9)	p[q] entails –(–p[q]) Co is p[q].–p[q], of which neither component is provable.		
D(1)	I		DR
	p[q]	⟶	–(–p[q])
Confirm p[q] by –p–q(i)	–p–q	TF implies	–p–q
D(II)	I		DR
	–p[q]	⟶	–(p[q])
Confirm –p[q] by p–q(ii)	p–q	TF implies	p–q

In each case the DR has a single form of confirmation and so we need only assure ourselves that at least one form of confirmation of an I TF implies the confirmation of its corresponding DR.

(3) Entailments cannot be manufactured by the 'strengthening' of the condition. 'If p then r' does not, as it does not in Stalnaker's theory, entail 'If p and q then r'. The confirmation of p[r] does not confirm pq[r] in any alternative way. For similar reasons p[q] does not entail p[q v r].[13] In giving expression to a conditional we vouch for the relevance of each of its elements. The arbitrary addition of elements to a true conditional is therefore illegitimate. Although disconfirmation of p[q v r] disconfirms p[q], p[q] may be fully supported

in the absence of substantive evidence for p[q v r], which must be confirmed by p–qr(i), pq–r(ii), and –p–q–r(iii). Moreover, p[q v r] entails p–q[r] (they have the same confirmations and disconfirmations), and given the transitivity of entailment, if p[q] entails p[q v r], it will also entail p–q[r]. But from if 'Cindy comes to the party, Dave will come' one cannot deduce that if Cindy comes to the party and Dave does not, then J. F. Kennedy is still alive.

We should observe, however, that although p[q] does not entail p[q v r], p.p[q] does entail q v r. p.p[q].–q.–r contains the normal inconsistency p.p[q].–q and is therefore inconsistent. But this does not result in one's being able to deduce r from p.p[q].–q, as one could if p[q] were to entail p[q v r]. In OC p.p[q].–q is an inconsistency and incapable of bearing entailment relations with other forms.

The Stalnaker/Lewis theory also prohibits entailment that relies on the strengthening of the condition, on the ground that if the denials of the element or elements that are added to the condition of the antecedent conditional to form the consequent conditional happen to imply assumptions necessary to the truth of the original conditional, the consequent conditional will become false. In chapter 1, although agreeing with the conclusion, I dissented from the reasoning, arguing that relevant elements, if not made explicit in the condition, must be assumed to hold if the conditional is to be considered true. For this reason, 'If I strike this match, it will light' does not entail 'If I strike this match, and if it is wet, it will light' any more than is the latter entailed by 'If I strike this match and it is dry, it will light'. Thus, the example chosen by Stalnaker does not invalidate the classical law that 'If p then q' entails 'If p and r then q' because that law, when properly applied to ordinary discourse, does not result in the entailment of 'If I strike this match and it is wet, it will light' by 'I strike this match it will light'. What makes the law invalid in the present theory is that, in accordance with OC, the confirmation of the antecedent must confirm the consequent, and it is this requirement that disqualifies the entailment of pq[r] by p[r] and makes it impossible to multiply entailments by adding irrelevant elements to the condition of the antecedent.

Stalnaker would argue that it is inconsistent to hold that the conditional is transitive (as I do in 2(2)) and also that entailment through the strengthening of the condition is invalid, because the entailment of 'If p and q then r' by 'If p then r' is a consequence of the acceptance of transitivity. The justification of this result in the Stalnaker system is as follows.[14] His axiom (a2), taking '>' as the sign of conditionality, is

$(\Box(A \supset B) \supset (A > B)$.

which may be expressed roughly in words as, "If A logically implies B, then 'If A then B' is true." If this is so, then from a true entailment, such as 'AC entails A', we may validly infer the corresponding conditional, 'AC>A'. But suppose, generally, that conditionality is transitive, so that any schema having the form

$(X>Y)(Y>Z)$

entails

$(X>Z)$

and that, in particular, (A>B). Then from the truth of (A>B), the truth of (AC>A) and the hypothetical syllogism, we may deduce (AC>B). By virtue of the operation of the application requirement, nothing like this happens in OC. The form, 'pr entails p' is true, but pr[p], far from being derivable from 'pr entails p', is not even properly formed. As just explained (see pp. 60–62), to participate in an entailment, a component must be capable in principle of being confirmed and also of being disconfirmed. If not, it does not fulfill the application requirement and is not interpretable by a statement. Therefore a conditional schema that would take only contradictory disconfirmation is not a proper construction. The form of the disconfirmation of pr[p] is a contradiction and therefore pr[p], not being well formed, cannot be entailed by 'pr entails p'. In OC the transitivity of the conditional is not incompatible with the rejection of the entailment of pr[q] by p[q].

VIII

OC is consistent in the sense that it is never a truth of the system that one formula both entails and does not entail another. Let us call the denial of an inconsistent conjunction a 'valid' formula, with the result that, for example, to the inconsistent conjunction A.B.C. there corresponds the valid derivation –A v –B v –C, which is the denial of A.B.C.; and to the inconsistent conjunction A.B v C.D there corresponds the truth functional equivalent of its denial, the valid formula (–A v –B)(–C v –D).

On the assumption that the forms H and G are contingent, H entails G if and only if –H v G is valid. Every entailment of G by H is equivalent to the validity of –H v G and for every valid structure –H v G, in which each of –H and G is contingent, there is a true entailment of the form 'H entails G'. If 'H entails G' is true, '–H v G is valid' is also true and so 'H.–G is inconsistent' is true. Then 'H does not entail G' is not also true, because if this were so, –H v G would not be valid and H.–G would not be inconsistent. And to say that a formula is and is not inconsistent is a contradiction.

'H entails G' is true if and only if '–G entails –H' is true because each says the same thing, that H.–G is inconsistent. Also, if H entails G and G entails J, then H entails J. The argument for this conclusion is as follows. Clearly, if the form A is inconsistent, then the form (A)(B v –B), which may be expressed as A.B v A.–B, is inconsistent, because each disjunct contains an inconsistency. And the converse is also true. If A.B v A.–B is inconsistent, the inconsistency of neither A.B nor A.–B can depend upon the truth or falsity of B; if it did, then one of A.B or A.–B would be consistent and the whole would not be inconsistent. Therefore A is inconsistent if and only if (A)(B v –B) is inconsistent. Now if H entails G and G entails J, then

(H.–G) v (G.–J)

is inconsistent and so also is

(H.–G.–J) v (H.G.–J).

But this last is the truth functional equivalent of

(H.–J(G v –G)

and because this is inconsistent, so is (H.–J). Thus entailment in OC is transitive.

The system embodies a decision procedure. It provides a method by which one can find out, with regard to any collection of statements of the type whose entailment relations are studied in the essay, whether the conjunction of that collection would be inconsistent, and, if the conjunction is inconsistent, the entailment relations that can be derived from it. The system does not, however, purport to be complete. OC attempts to present the logic of a province of natural language. It cannot be shown that all and only the entailments demonstrable in OC are all and only the true entailments obtaining in that province of the language. I will argue in chapter 6 that we cannot preclude the possibility that one may be in a position to say that to an instance which appears to be an entailment in natural language there is no corresponding entailment in OC or that to an entailment demonstrable in OC there is an exception in natural language. If there existed some system X that was known to be perfect, we might be able to discover whether OC corresponds to X. But knowledge of the perfection of a system is not to be had—X, as much as OC, must face the tests with which it is confronted by natural speech, which is an evolving phenomenon without natural boundaries.

OC is a theory of the entailment relations of a limited assortment of types of statement. This chapter has been concerned with singular nonconditional and ordinary conditional statements and in chapter 3 I examine monadic nonconditional and conditional general statements. Deductive relations of certain types of conventional and logical statements are presented in chapters 4 and 6. I have noted that polyadic general statements and also various types of conditional statements—such as 'even if' and disposition conditionals—fall outside the scope of the theory. Also, not all the statements that form part of the theory can take places on an entailment line. Singular statements on the line are assumed to hold during the single occasion to which they have reference. Phase statements that form parts of the confirmations of conditionals refer to segments only of the occasion in question and do not qualify for positions on the entailment line. Their relations however, to the statements that appear on that line are fully defined.

I do not propose to indicate in this essay, nor do I know, how the many additional types of factual and other statements, such as polyadic general statements, might be incorporated in the OC system. And as stated in the Preface, I am not able to say how a standard system, for example, a natural deduction system, could be modified to absorb the theory of conditionality contained in OC.

Chapter 3

GENERAL CONDITIONALS

I

We use conditionals of the general form 'If a thing is f, then that thing is g', or 'Necessarily, if a thing is f, it's g', to express the thought that, because of the relations things bear to each other, a thing's being f ensures its being g. It is the form we most frequently use to propound the relations we believe to exist in nature between things and we continue to use it in spite of the fact that we know that the truth of any assertion having the form is not deducible from the truth of any nonconditional assertions about the properties of things. This unrestricted conditional has a different logic from, although it has the same visible form as, the conditional by which we state that all members of a finite sample of things f, (for example, the marbles in this bag) have a certain characteristic. 'All the marbles in this bag are blue' is not an ordinary conditional. It can be established to be true by inspection of the marbles, just as 'The legs of this chair are made of wood' can be established to be true by examination of the legs. The unrestricted conditional 'All things that are f are g', however, is used to say that all things past, present and future, because of permanent arrangements holding among things in the world, are g if they are f. Statements about finite collections and unrestricted conditionals, although they may be similar in appearance, have different functions: the one states that members of a finite group have some feature in common; the other that all things f and g are related in such a way that a thing's being f is ground of its being g.

No one disputes that the logical properties of this unrestricted or universal conditional diverge from those of the universal generalization of the material conditional of standard predicate logic. 'Always, if it's an f then it's g' is not the logical equivalent of "For all x, either x is not-f or it's g'. In this essay I treat 'A entails B' as meaning that A and B have properties which make B formally deducible from A. This being the case,

I(1) $-(\exists x)(fx)$ entails $(x)(fx \supset gx)$,
I(2) $(x)(gx)$ entails $(x)(fx \supset gx)$, and
I(3) $(x)(fx \supset gx).(x)(fx \supset -gx)$ is self-consistant

are all true. But if we use 'Always, if it's f then it's g' in the sense that implies that a thing's being f is always ground for, and justifies, the expectation that it is g, then we regard

II(1) 'Everything is –f' entails 'Always, if a thing is f then it's g,'
II(2) 'Everything is g' entails 'Always, if a thing is f then it's g' and
II(3) 'Always, if a thing is f, then it's g and if a thing is f then it's –g' is self–consistent

as all false. Logicians disagree on the questions of the importance of these differences and as to whether they are more apparent than real. They also disagree as to the reasons for the differences. Some claim that the unrestricted conditional is true, when it is true, because events of the type described by the condition cause events of the type described by the conclusion. Others argue that we have no grounds, either empirical or logical, for the belief in a relation between events of causal necessitation and that the differences in our treatments of $(x)(fx \supset gx)$ and 'Always, if it's f then it's g' are founded only in differences in attitude. But in spite of these disagreements, most writers do concur that the two types of conditional differ at least in the ways just illustrated. They would also agree, I think, that part of what we claim in asserting 'Always, if a thing is f, then it's g' is that if a thing is f, then because of the way things are ordered in the world, its being g is also ensured. And if it is true that 'Always, if a thing is f then it's g' at least means this, then it would seem possible to regard it as a universalization of the singular conditional that formed the subject matter of chapter 2. The individual thing or occasion that is the common reference of the condition and conclusion of the singular conditional is replaced by a variable that is bound by the operator 'always' or 'necessarily': 'Always (necessarily) if it's invertebrate then it's cold blooded'. The particular conditional is created when the variable is bound by 'sometimes' or 'possibly'. I use 'always' or 'necessarily' to convey the notion that it is in the nature of the case, inevitable, that things have a certain characteristic, and 'sometimes' or 'possibly' the notion that a thing might have a characteristic or that it is not inevitable that it not have the characteristic.

In this chapter I shall show that a logic of general conditionals based on this conception produces entailments that express rather well the ordinary convictions we have about the logical relations between general conditionals and between general conditionals and nonconditional statements. This logic preserves, for example, the validity of contraposition and transitivity for universal conditionals, and we are able to show that 'Necessarily if it's f then it's g' entails 'If it's f, it's possibly g' or 'If it's f it does not on that account follow that it's not g'. An interesting result of this theory of general conditionals is that, on a certain interpretation of the 'All' and 'Some' propositions of traditional logic, all the valid forms of the syllogism and the relations of the square of opposition are provable, without the aid of existence assumptions, as theorems of the system OC. Also, no formulation corresponding to an invalid form of the syllogism is logically true in the system. The chapter will be for the most part constructive and I will not attempt a review of the many theories of the general conditional that have been offered as alternatives to the standard theory of conditionality.

It will have to be borne in mind that the words 'necessarily' and 'possibly' will not have the same import, respectively, as the quantifiers '(x)' and '(∃x)' of standard formal logic. Statements using these latter expressions convey something about existence. The form, '(∃x)(hx)' says that there is something that is h and '(x)(hx)' says that there is nothing that is not h. The form, '(x)(gx ⊃ hx)' states that nothing is g and not h, whereas its denial '(∃ x)(gx.–hx)' claims that there is something that is g but not h. 'Necessarily' and 'possibly' in OC, however, make no direct claims about the existence or nonexistence of things of certain types. 'Possibly h' in OC states that things are not arranged in nature in such a way as to preclude the occurrence of things that are h, whereas 'Necessarily h' asserts that natural relations *are* such as to preclude the occurrence of things that are not h. 'Necessarily, if a thing is g than it's h' says that gs and hs are related in such a way that anything that is g is h. And the denial of this assertion, which I shall express by 'Possibly if a thing is g then it's not h,' denies the relation. It says that if a thing is g, one may not on that account conclude that it is h. It does not claim that there are things that are g and not h.

It may be objected that 'Always, or for all x, if gx then hx' need not be taken as equivalent to the denial of the existential form '(∃x)(gx.–hx)' but as the assertion of the conditional relation between g and h. Yet for our purposes, because of the familiar association between '(x)' and '(∃x)', the principal drawback of rendering the universalization of the conditional as '(x)(if gx then hx)' lies in the consequences of its denial. It is true that 'It is false that for all x, if x is g then it's h' or 'Not all gs are hs' may be understood as denials of a relationship between gs and hs and not as assertions that there are gs that are not hs; but they are regularly read as 'Some gs are not hs' and this reading does push us toward the conclusion that there are things that are g but not h. The use of 'all' inclines us to make the mistake of assuming that the denial of a universal conditional has existential implications. The advantage of the modal word 'necessarily', however, is that its denial goes with 'possibly not': 'It is not necessarily the case that if a thing is g, then it's h' is most naturally read as 'Possibly if a thing is g, it's not h' and this latter statement more accurately expresses the nonexistential claim of the denial of the universal conditional, which is that, if a thing is g, it does not on that account follow that it is h. The existential neutrality of the pair 'necessarily' and 'possibly' is useful not only in the development of the logic of general conditionals but also in the applications of that logic to the traditional system of the syllogism (as will be seen in sec. 7 of this chapter).

Of course 'Necessarily if a thing is g, it's h' and 'Possibly if a thing is g, it's not h', although not implying the existence of things that are g and h or that are g and not h, do have logical relations with existential statements. Thus if there exists something that is both g and not h, then 'Necessarily if a thing is g then it's h' is disproved and its denial is proved. But I will not here examine in a systematic way the relations between the operators 'Necessarily' and '(x)' and 'Possibly' and '(∃ x)'.

We must guard against the temptation to treat 'Necessarily' and 'Possibly' here as expressive of logical relations. 'Necessarily if a thing is g then it's h' does not assert that if a thing is g then its being h follows as a matter of logical necessity. The necessity claimed by this statement is not logical but in nature. We should also resist, I suggest, the impulse to think of natural necessity exclusively in terms of causal necessity. As we saw in chap. 2, the connection that justifies the assertion of a conditional relation may be a purposive social or psychological one, not obviously reducible to a causal link. And just as there are true singular conditionals flowing from purposive relations, so there may be true universal ones. Thus the association that entitles us to conclude to g on the occurrence of f need not (as a matter of logic) be a causal one.

Before proceeding further, it is important to notice that the universal conditional is unprovable in two senses. It is unprovable, first, because each of its singular exemplifications is unprovable. 'Necessarily all mammoths are monogamous' implies, 'If this is a mammoth it's monogamous.' But from the observation that this is a mammoth and monogamous, we cannot infer, on logical grounds, that the fact that it is a mammoth is a sufficient justification for concluding that it is monogamous, any more than we can infer, for reasons of logic, that if this is a pen then it's black, from the fact that this is a pen and black. The universal conditional is unprovable, secondly, in the sense that if the best possible direct evidence for its truth is afforded by all exemplifications of it so far adduced, it remains possible that some further case will prove unfavorable.

The three ideas that governed the logic of the singular conditional will also guide the development of the logic of the general one. Logical relations between factual assertions, conditional and nonconditional, are determined by the associations of their confirmations and disconfirmations. Also, if A is to entail B, that is, if A.–B is to be inconsistent, then each conjunction in the disjunction of conjunctions formable from A.–B must contain a normal inconsistency, which is to say, an inconsistency such that the confirmation of each I it contains confirms its corresponding DR and the disproof of each disprovable DR disproves I. Finally, the subject matter of the logic will be the logical relations of factual statements, and to rank as factual a form must be formally contingent. Only contingent forms may qualify for positions in the antecedent and consequent of an entailment.

II

Let us now apply the concepts of confirmation and disconfirmation to the general form. The letters 'a.b.c . . . f.g.h . . .' will stand for predicative phrases and 'N' and 'M', subject to these qualifications, will stand for 'Necessarily' and 'Possibly', respectively. As was the case for singular conditionals, I use here an unfamiliar method of representing the form of general conditionals so that the reader will not be tempted to impute to the general conditionals of OC characteristics associated with conditionals expressed by other symbolisms. The form a[b(N)] is read 'Necessarily, if a thing is a, then that thing is b'. The part of a[b(N)] which lies outside the square bracket is the subject matter of the condition of the form and the predicative expression inside the square bracket stands for the subject matter of the conclusion. The form a[b(M)] is read 'Possibly if a thing is a, that thing is b' or 'If a thing is a, it cannot on that account be concluded that that thing is not b'. (I will sometimes express 'Necessarily if a thing is a, that thing is b' and 'Possibly if a thing is a, that thing is b' by, respectively, 'If a thing is a, it's necessarily b' and 'If a thing is a, it's possibly b', meaning no more by the latter pair than by the former—the necessity or possibility, if it obtains, is in the relation between the events, not in the events themselves.) The operators 'N' and 'M' bind the expressions inside and outside the square brackets. Nonconditional forms are those that lack conditions and have conclusions only. 'Things are necessarily a' or 'One is entitled to say of everything that it is a' has the form a(N) and 'Things are possibly a' or 'One is not entitled to conclude of everything that it is not a' has the form 'a(M)'. To say that things are possibly a or that a is possible is not to say that if no individual exemplifies a in the actual world, a is being realized in a possible although unactual one. The form, a(M) says that some individual may instantiate the event or state a in the actual world. If something has a, a(M) is proved; but no collection of negative instances will disprove it. On the other hand, a(N) is disproved by one negative instance but cannot be proved by any number of positive instances of a.

'Components' of the system are forms such as a[b(N)], ab[c(M)], a v b(M) . . . whose subject matters are bound by a single operator. The logical constants 'v,' '.,' and '–', which connect forms or that connect confirmations and disconfirmations will continue to have the logical functions they have in truth functional logic. Two forms placed side by side will be read as conjoined.

A component containing one operator may be converted into the equivalent form containing the other operator by denying the whole form, denying the subject matter of the conclusion and changing the operator. (I indicate the denial of the 'whole form' here by placing the negation sign outside brackets that contain the expression to be denied. The form, –(–a(M)), for example, says that it is not the case that –a is possible. It does not assert the different statement that – –a, or a, is possible. Thus a[b(N)] is equivalent to –(a[–b(M)]) —in words, 'If a thing is a it's necessarily b' is equivalent to

'It's false that if a thing is a, it's possibly not b'. The form a[b(M)] is equivalent to –(a[–b(N)]), a(N) to –(–a(M)) and a(M) to –(–a(N)). Because of the interpretation I have given to M, a(M) has to do not with the actual existence of things that are a, but with the nonlogical possibility of there being such things, or more precisely, with the absence of a warrant to conclude of everything that it is not a. Also, a[b(M)] does not imply that things are possibly a and b. I will show later that a[b(M)] does not entail ab(M), and that a[b(M)] is consistent with –a(N). The truth of 'If a high jumper has cleared twenty feet, there is no assurance that on that account he is not blue eyed' (or, differently expressed, 'If a high jumper has cleared twenty feet he is possibly blue eyed') does not give us the logical right to say that there possibly exist blue–eyed high jumpers who have cleared twenty feet, because clearing twenty feet may not be humanly possible.

The calculus of general statements offered here is a monadic predicate calculus, which is to say that the expressions governed by a modal operator 'necessarily' or 'possibly', for instance, a[b(N)], b v c(M) . . . are to be attached to one individual at a time. Thus, for example, the condition and conclusion clause of a[b(N)] are to be instanced by the same individual.

Universals refer not to occurrences of particular events but to all occurrences. They may be instantiated at any time by any occurrences of the events they describe, and so we may take as exemplification of a universal an occurrence during any occasion or segment of an occasion. Let f1, f2, . . . express the facts that individuals 1, 2, . . . have the quality described by f during certain occasions. If we wish to express the fact that the individual 1 has the characteristic f during a segment of an occasion and, say, –f during another segment, the notation f1(i), –f1(ii) will be used. Confirmation of the form a[b(N)] consisting of a1b1(i), –a1–b1(ii) . . . anbn(i), –an–bn(ii) is the type of support obtained when one is able to perform an experiment—to introduce, or observe the introduction of, changes in the status of an individual and record concomitant changes that may occur. I will not complicate the exposition with considerations of this kind, however, and will treat illustrative instances of generalities as, unless the contrary is indicated, situations, expressed by f1, f2, . . . and so on, in which an individual, for instance, 1, exhibits a feature f throughout the occasion. Nevertheless, for the purposes of determining the logical relations between statements, the logical possibility of the division of confirming occasions into phases will be assumed, and this potential feature of the confirmation of general conditionals will be exploited in the presentation of the relations between a singular conditional and its corresponding universal (see pp. 98–99).

Confirmations and disconfirmations will consist of truth functional compounds of forms of this type. Let a,b,c . . . f,g,h . . . stand for simple or complex predicates. The form f(N) requires that f be true of everything. Therefore the confirmation of f(N) will consist of an indefinitely long conjunction,

f1.f2.f3 . . .

The form, f(N) is disconfirmed if it is found that f is true of any individual. The disconfirmation of f(N) is then the indefinitely long disjunction,

–f1 v –f2 v –f3 v . . .

The confirmation of the denial of f(N), –(f(N)), is the disconfirmation of its assertion and the disconfirmation of –(f(N)) is the confirmation of f(N). The confirmation of f(M) is the confirmation of its equivalent, –(–f(N)), and is therefore the disconfirmation of –f(N).

f1 v f2 v f3 v . . .

Because no finite set of instances implies an indefinite conjunction, whereas every finite set implies an indefinite disjunction, we say that the confirmations of N forms and of the denials of M forms do not prove them; but these forms are disproved by their disconfirmations. The confirmations of M forms and of the denials of N forms prove them, but their disconfirmations do not disprove them.

The claim that a statement is true is not also a claim that it has been confirmed. And the form of a statement, as it appears on the entailment line, does not imply the truth of its confirming statements. The setting out of the confirmation underneath the entailment line simply shows the form that confirmation would take were it to occur.

III

The universal conditional f[g(N)] is disconfirmed if the subject matter of the condition f and the denial of the subject matter of the conclusion, –g, are true of the same individual. The disconfirmation of f[g(N)] therefore has the form,

(f–g)1 v (f–g)2 v (f–g)3 v . . .

Notice that the individual numbers 1, 2, 3 . . . are placed outside the brackets for convenience only. If we like we may distribute them throughout the predicative expressions themselves. For example, f1.g1 and (fg)1 say the same thing.

The confirmation of the general conditional cannot be considered, any more than can that of the singular conditional, as the simple denial of its disconfirmation. If that view were taken, then f[g(N)] would be confirmed by (–f v g)1.(–f v g)2, . . . and this would have the result, leading back to the famous paradoxes, that f[g(N)] would be confirmed by an instance of –f v g. My proposal is that we treat as a confirmation of a universal conditional, in conformity with the practice established for singular conditionals, only a form that arises when one is bound to conclude (to save the truth of the conditional) to the presence of a feature by reason of the coming about of a threat of disconfirmation. Such an exposure is brought about for a component P if all but one of the elements in a conjunction, realization of which would disprove p, are satisfied. If the remaining element of the disconfirmation of the form is realized, P is disconfirmed; if the denial of the remaining element is fulfilled, P is confirmed. Then f[g(N)] is placed in jeopardy by a fulfillment of f or –g. The confirmation of f[g(N)] has any form that would result from the making of a risk avoiding response to an endangering situation: an instance of (fg) or of (–f–g). If the available confirmations consist of instances of both forms, the component is said to have been confirmed in each alternative way. Such a collection would be a conjunction of the form,

(fg)1.(fg)2 . . . (–f–g)n . . .

It is not being proposed that, if f[g(N)] is true, every individual has the characteristics expressed by fg or –f–g. It is rather that, in collecting evidence for f[g(N)], we put together groups of this type, ignoring cases of –fg as lacking confirming force. If our investigations uncover a case of f–g, f[g(N)] will of course be disproved.

The form, fg[h(N)], with disconfirmation form fg–h, is threatened by the exemplification of fg, g–h, or f–h. It is confirmed in each alternative way if confirmed by a collection including instances of fgh, –fg–h, and f–g–h. Single disconfirmations may always be restated in disjunctive form, as disjunctions of one or more conjunctions, and a conditional receives single confirmation, is singly confirmed, if a confirming response has been obtained for the case of each conjunction in the disjunction forming its disconfirmation. The form of disconfirmation of f[gh(N)] is (f)(–g v –h), which has the disjunctive normal form (f–g) v (f–h). Any single confirmation will therefore consist of a response to the danger of realization of (f–g) and one to the danger of fulfillment of (f–h). A single *full* confirmation will contain an instance of each alternative response to the threat of satisfaction of these forms. For f[gh(A)], an instance of f threatens to fulfill both (f–g) and (f–h) and therefore, when placed in a state of risk by f1, f[gh(N)] is confirmed by (fgh)1, which is one of the alternative confirmations of f[gh(N)]. The form, (f–g) is almost fulfilled by an instance of –g and (f–h) by an instance of –h; therefore, the remaining alternative confirmations of f[gh(N)] would be (–f–g)2 and (–f–h)3. Thus, the full confirmation of f[gh(N)] would involve instances of (fgh), (–f–g), and (–f–h).

The only direct evidence that can be obtained for the truth of a conditional of the universal form f[g(N)] is, first, the negative evidence that it has not been disconfirmed, and secondly, the positive evidence consisting of fulfillments of its confirmation forms (fg) and (–f–g). As was the case for the singular conditional, although an instance of (fg) and an instance of (–f–g) are each called 'confirmations' of f[g(N)], only the confirmation of the universal conditional in each alternative way adds significantly to its justification—provides substantive support for the claim that the fact of a thing's being f entitles one to conclude that it is g.

What confirms the denial of a universal conditional is what disconfirms its assertion. And because the rule has been adopted that a general statement can be transformed into its equivalent containing the other operator by denying the statement, denying the subject matter of its result and changing the operator, then –(f[–g(N)]) is the equivalent of f[g(M)]. Therefore, their confirmations and disconfirmations are the same. The evidence in favor of f[g(M)] will have the form,

(fg)1 v (fg)2 v (fg)3 v . . .

and the evidence against it will have the form,

(f–g)1.(f–g)2.(f–g)3 . . . (–fg)n . . .

The confirmation of f[g(M)] proves it at both levels. Thus (fg)1 proves, with regard to the individual instance 1, that this is not an instance of which 'If f then –g' is true; it also proves the general claim that if a thing is f, it cannot on that account be concluded that it is not g. No finite body of disconfirming evidence disproves a particular conditional, however, but only makes its truth less likely.

A general conditional, universal or particular, is tested by the instantiation of any group of its predicate expressions that consists of all but one of the predicates contained in a conjunction that constitutes a disproof or proof of the conditional but does not imply the truth or falsity of the instantiation of the remaining predicate. The universal conditional is therefore tested by its risk factor; as a result of the bringing about of the endangering situation, the assertion will either be confirmed or disproved. A particular conditional is tested by the instantiation of all but one of the predicates contained in a conjunction that would constitute its proof, and the result of this exemplification will be either the proof or disconfirmation of the conditional. The testing of the particular conditional is not, however, crucial because the particular conditional is not capable of disproof. Thus to place a proposition in jeopardy is to test it, but to test it is not necessarily to place it in jeopardy.

It is true of general expressions, as it is of singular ones, that a component must be contingent if it is to play a role in entailment. This means, with regard to nonconditional components, which possess only conclusion clauses, that their conclusions must be neither tautologous nor contradictory. With regard to conditionals and their denials it means that test factors must not truth functionally imply either the affirmation or denial of the fulfillment of the remaining predicate in the disconfirming conjunction containing the test factor. Such an eventuality would make it impossible to 'test' the form in one or more of the ways inherent in the form itself and would result in its disqualification. For example, the form –f[f(N)] is unacceptable because its risk factor, –f, . . . truth functionally implies the remaining predicate in the disconfirmation of the conditional, which has the form –f . . . –f . . . . The form fg[f(N)] also infringes the rule: f . . . g . . . TF implies f, . . . g . . . –f . . . TF implies –f . . . and f . . . –f . . . is a contradiction. It is when a conditional expression possesses a test factor not capable of creating an actual test that it fails to qualify as a usable expression—it must be regarded as defective in form.

Nonconditionals do not, as conditionals do, assert that the predicates they contain have reference to things related in nature. Therefore, in seeking confirmation of nonconditionals we do not need to look for or bring about testing situations the results of which will support or confirm the nonconditional statement. The nonconditional, one might say, is tested by any situation; its condition clause is 'empty' and in justifying it we need only concern ourselves with exemplifications of its conclusion clause. The particular nonconditional is proved if its predicate is true of any proffered thing; the universal nonconditional must be true of all things.

The establishment of entailments between general statements is not different from that of entailments between singular statements, except that instead of being limited to occurrence during a single occasion, we may gather confirmations and disconfirmations from all occasions during which the predicates, or the negations of the predicates, forming the subject matter of the general statements, are instanced. The general form A entails the form B if A.–B is inconsistent (where A and B are contingent), and A.–B is inconsistent if each conjunction in its disjunctive normal form, C1 v C2 v C3 v . . . contains a normal inconsistency, namely a conjunction (1) which is such that the confirmation of each of its provable components TF implies the disproof of the other components and (2) is such that when the implicans has been confirmed in each alternative way, the denial of the remainder (DR) will have been confirmed in each alternative way.

Forcing. Because they are disproved by their disconfirmations, universal components, conditional and nonconditional, of an implicans are 'forced' by the confirmations of other components in the conjunction—the nondisconfirmation of the disprovable components must be fulfilled. Particular components are not subject to forcing, although their own confirmations may force the components with which they are conjoined. For a discussion of the mechanism of 'forcing,' see chap. 2, (pp. 70–71).

IV

A confirmation of a universal nonconditional consists in the predication of its subject matter of an individual and its disconfirmation is the denial of such predication. It is therefore not necessary, in establishing the inconsistency of a conjunction consisting wholly of nonconditionals, to show that the confirmation of each implicans TF implies its DR. One test will do; because the denial of the confirmation of a nonconditional is the confirmation of the denial of the nonconditional. Thus if the nonconditional implicans A has the appropriate relation to its DR, B, so will the implicans –B to –A. (In Theorem 3(1), however, I set out each I and DR in order (see p. 95.) One must nevertheless observe the restriction that requires that the proof of a component of Co must disprove the rest of the conjunction. The form, g(M) does not entail g(N) because in g(M).–g(M) the proof of neither component disproves its associate. In 3(1), however, we see that the converse relation is an entailment. We again mark the different Is by D(I), D(II), . . . and indicate on the left the means by which the Is are to be confirmed.

3(1) g(N) entails g(M)
Co is g(N).–g(N)

Co passes the first test on inconsistency because neither g(N) nor –g(N) is provable.

D(I)	I		DR
	g(N)	⟶	g(M)
Confirm g(N) by g1	g1	TF implies	g1

D(II)	I		DR
	–g(N)	⟶	–g(M)
Confirm –g(N) by –g2	–g2	TF implies	–g2

The confirmation of each I confirms the denial of the remainder. The proofs of the following entailments will help to illustrate the way in which the system applies to particular and universal forms. Recall that if the denial of the remainder is an M form or a nonconditional general form, it will have only one alternative mode of confirmation, and it is sufficient if we can find one form of confirmation of the implicans that TF implies a DR of that type. Remember also that, if one component of an I is confirmed, its disprovable associates must not be disconfirmed: the denials of their disconfirmations must be true.

3(2) g(N).g[h(N)] entails h(N)
Co is g(N).g[h(N)].–h(M)

The proof of Co's only provable component, –h(M), would result in the disproof of the rest of Co: –h1 TF implies –g1 v g1–h1.

D(I)	I		DR
	g(N).g[h(N)]	⟶	h(N)
Confirm g(N) by g1	(g1)(–g1 v h1)	TF implies	h1

D(II)	I		DR
	g[h(N)].–h(M)	⟶	–g(M)
Confirm –h(M) by –h2	(–g2 v h2)(–h2)	TF implies	–g2

D(III)	I		DR
	g(N).–h(M)	⟶	g[–h(M)]
Confirm –h(M) by –h3	(g3)(–h3)	TF implies	(g3–h3)

Notice that in D(II) and D(III) one selects –h(M) for confirmation because it is not itself capable of being forced but its confirmation forces avoidance of disconfirmation on the part of the (disprovable) universal with which it is conjoined.

3(3) f[g(N)] entails f[g(M)]
Co is f[g(N)].f[–g(N)]

Neither f[g(N))] nor f[–g(N)] is provable and so Co passes the first test on inconsistency.

D(I)	I		DR
	f[g(N)]	⟶	f[g(M)]
Confirm f[g(N)] by f1g1	f1g1	TF implies	f1g1

D(II)	I		DR
	f[–g(N)]	⟶	f[–g(M)]
Confirm f[–g(N)] by f2–g2	f2–g2	TF implies	f2–g2

Observe the similarity between 3(3) and the inference called 'subalternation' in traditional formal logic: if all A is B then some A is B. When the antecedent and consequent are reduced, in the classical way, to 'Nothing is both A and not–B' and 'Something is both A and B', respectively, subalternation becomes invalid because the truth of 'Nothing is both A and not–B' does not assure us of the truth of 'Something is both A and B.' In the present system, however, f[g(M)] says only that if a thing is f, one cannot on that ground conclude that it is not g, and this does not entail that there exists something that is both f and g. The form, f[g(M)] does not even entail the modal form fg(M), which says that something's being f and g is possible. The Co of a putative entailment between f[g(M)] and fg(M) would be,

f[g(M)].(–f v –g(N)).

The confirmation of (–f v –g(N)), –f1 v –g1, does not TF imply either f1–g1 or –f1g1, the alternative confirmation of the DR, f[–g(N)]. But fg(M) does entail f[g(M)]:

3(4)

fg(M) entails f[g(M)]
Co is fg(M).f[–g(N)]

The only provable component of Co is fg(M), and its proof disproves f[–g(N)].

D(I)	I		DR
	fg(M)	⟶	f[g(M)]
Confirm fg(M) by f1g1	f1g1	TF implies	f1g1

D(II)	I		DR
	f[–g(N)]	⟶	–f v –g(N)
Confirm f[–g(N)] by f2–g2	f2–g2	TF implies	–f2 v –g2

The universal conditional entails the universal material conditional.

3(5)

f[g(N)] entails –f v g(N)
Co is f[g(N)].f–g(M)

Proof of the only provable component, f–g[(M), disproves the conjunct f[g(N)].

D(I)	I		DR
	f[g(N)]	⟶	–f v g(N)
Confirm f[g(N)] by f1g1	f1g1	TF implies	–f1 v g1

D(II)	I		DR
	f–g(M)	⟶	f[–g(M)]
Confirm f–g(M) by f2–g2	f2–g2	TF implies	f2–g2

But the converse relation is not an entailment. The Co of that relation is

–f v g(N).f[–g(M)],

and the confirmation of –f v g(N) does not TF imply either of the confirmations of the denial of f[–g(M)], which is f[g(N)]. The reader may wish to satisfy himself that the universal conditional is transitive and that it entails its contrapositive:

3(6) g[h(N)].f[g(N)] entails f[h(N)][1]

3(7) f[g(N)] entails –g[–f(N)]

In combining general and singular forms in deductions, we must keep in mind that the universal form is required to be true of everything, whereas the particular need only be possibly true of one or more things. If the universal h(N) is true, each of h1, h2, . . . is true and so in assessing the entailment relations of h(N) to a singular statement formed from the instantiation of a predicate by 1, we consider the effect of confirming h(N) by h1. The particular form h(M), however, is capable of proof by the instantiation of h by any individual, and so in assessing its entailment relations to singular form, we consider the effect of its instantiation by any individual other than the one to which the singular form applies. With respect to singular and general conditionals, we must remember that the proper confirmation of singular conditionals is to be recorded during the phases of the occasions to which they refer. In reviewing the consequences for the truth or falsity of a singular conditional, of the truth of a universal conditional, we consider what the effect would be of the phased confirmation of the universal upon the phased confirmation of the singular.

The form h1 entails h(M). Their Co is h1.–h(N). The form, –h(N) is not provable but is disproved by the proof of h1. Also, the confirmation of –h(N) by –h1 TF implies the confirmation of the denial of h1 and the confirmation of h1, as noted, TF implies the confirmation of the denial of –h(N), that is, h(M). But h(M) does not entail h1. The proof of –h1 does not disprove h(M). The universal conditional entails the singular:

3(8) a[b(N)] entails a1[b1]
Co is a [b(N)].–(a1[b1])

The proof of –(a1[b1]) by a1–b1 disproves a[b(N)] and the latter is not provable. The form, a[b(N)] refers to all individuals and '1' identifies a particular individual, not a phase of an occasion. Note, however, that the confirmation of both a[b(N)] and a1[b1] is divisible into phases, as was observed at p. 89.

D(I)	I		DR
	a[b(N)]	⟶	a1[b1]
Confirm a[b(N)] during a phase			

of the occasion to which a1[b1] applies by a1b1.	a1b1(i)	TF implies	a1b1(i)
Confirm a[b(N)] during phase (ii) of the occasion by –a1–b1	–a1–b1(ii)	TF implies	–a1–b1(ii)
D(II)	I		DR
	–(a1[b1])	⟶	a[–b(M)]
Confirm –(a1[b1]) on the occasion to which it applies by a1–b1 or at any phase	a1–b1	TF implies	a1–b1

The singular conditional does not entail the universal. The conjunction of the affirmation of the singular and denial of the universal is,

a1[b1].a[–b(M)].

It is possible that the proof of a[–b(M)] will be realized on an occasion other than the one to which a1[b1] applies and will not disprove a1[b1].

V

The universal conditional, like the singular, is not true unless each part of its condition is relevant to its conclusion: unless, given normal background conditions, the possession by a thing of each of the characteristics described by a conjunctive condition is relevant to the existence of the state of affairs described by the conclusion. One way the logic of OC expresses this feature of the ordinary conditional is in the theory's requirement that every characteristic mentioned in the condition or conclusion have a real function or be of real consequence. The theory does not permit, for example, the multiplication of entailments by the addition of sheltered elements to condition or conclusion. Thus to the fact about singular conditionals that p[q] does not entail pr[q] or p[q v r], there corresponds the fact about universal conditionals that a[b(N)] does not entail ac[b(N)] or a[b v c(N)].

This feature of the theory imposes a stringent demand on entailment between conditionals; but it is one I suggest we respect in our efforts to state relationships between things and events. This is particularly true in scientific discourse, where irrelevant additions are painstakingly excluded from law statements. In fact, part of the point of systematic experimentation is the elimination of features of a situation not relevant to the natural relations under study. But it is also true of our ordinary communications. If I have good grounds for asserting a[b(N)], I will have accumulated many cases of a . . . b . . . and of –a . . . –b . . . and I will have come across no cases of a . . . –b . . . . But if this is so, I am not justified in asserting a[b v c(N)]; because one thing that the latter says is that if a . . . –b . . . is instantiated, then c . . . will be instantiated—and although this claim *may* be true, evidence of its truth will not be forthcoming if a[b)N)] is true.

I observed not long ago an advertisement for instant coffee that said, in effect, 'Every tenth jar of our coffee contains a coupon worth one dollar or one hundred dollars'. Now, by the terms of standard predicate logic, if (A) 'Every tenth jar of our coffee contains a coupon worth one dollar' is true, it follows that (B) 'Every tenth jar of our coffee contains a coupon worth one dollar or one hundred dollars' is also true. But although disproof of the claim (B) would disprove (A), still, it is clear that if, after many inspections, we found that each examined tenth jar contained a coupon worth one dollar but that none contained a one-hundred-dollar coupon, we would accept (A) and reject (B). If the manufacturer were to defend himself by saying that (A) is true and, because (A) entails (B), (B) is true, we would respond that (A) does not entail (B). The consumers' objection corresponds to the failure, in OC, of the entailment of a[b v c(N)] by a[b(N)]. The Co of the putative entailment would be a[b(N)].a[–b–c(M)]. The confirmations of a[b(N)] do not TF imply the confirmations of a[b v c(N)], which are of the form a–bc, ab–c, and –a–b–c.

VI

It must be said that there are entailments in standard logic that have not been thought to be odd in any way but that are not entailments in OC. For example,

> (1) 'Necessarily if a thing is f then it's g and necessarily if a thing is both g and h, then it's j' entails 'Necessarily if a thing is both f and h, then it's j'

has long been considered a truth[2] and, when conditionality is represented by '⊃' as in the predicate calculus, it *is* a truth. The form (x)(fx ⊃ gx).(x)(gxhx ⊃ jx) entails (x)(fxhx ⊃ jx) because the conjunction formed from the assertion of the former and denial of the latter produces a contradiction. But it is a fact about this relation that, when its components are taken as conditionals, one may completely confirm 'If f then necessarily g' and 'If g and h then necessarily j', in each alternative way, without confirming 'If f and h then necessarily j' in each alternative way; and for this reason (1) must be rejected. 'If f then necessarily g' may be completely confirmed without forcing 'If g and h then necessarily j' to yield confirmation of 'If f and h then necessarily j' by f1h1j1. If 'If g and h then necessarily j' is confirmed by –g2h2–j2, we have (–f2 v g2)(–g2h2–j2), which yields –f2h2–j2, one of the confirmations of the proposed consequent. But neither of the other confirmations of the consequent, f1h1j1 and f3–h3–j3, is forthcoming from the testing of the antecedent.

The failure of this relation to support an entailment can be justified. Suppose, for example, that 'Necessarily if a thing is f then it's g' is interpreted by (i) 'Necessarily if a thing is a rattler than it's a cross-banded snake' and 'Necessarily if a thing is both g and h then it's j' by (ii) 'Necessarily if a thing is a cross-banded snake and indigenous to this area, then it's harmless'. Our confidence in (ii) may be based partly on the fact that no rattlers have been observed in the area and it is believed that they are not indigenous to it. But are we quite certain that if rattlers were indigenous to the area, they would be harmless? No matter how thoroughly we have tested (i) and (ii), it seems to me that if we learn that rattlers have taken up residence in these environs, we would want further tests to be made before we would be content to say that the indigenous rattlers are harmless. We would be loath to propound (iii) 'Necessarily, if a thing is a rattler and indigenous to this area then it's harmless' as a logical deduction from (i) and (ii).

Although I believe that the foregoing argument is correct, I expect that the reader will find the result unusual and perhaps difficult to accept. Because of the importance of the principle involved, I will show its application to the other main conditionals to which the essay has reference: The 'meaning' conditional or convention (see chapter 4, p. 133) and the logical conditional (chapter 6, p. 179–80).

VII

The difficulties of expressing the traditional logic of terms with the tools of modern predicate logic have been the subject of many discussions by text writers in recent times.[3] And given the treatment accorded by modern logicians to the four traditional forms of general statement, these difficulties are certainly inescapable. If A and E forms, 'All S is P' and 'No S is P', are understood to have the same logic, respectively, as 'For all x, if x is S, x is P' and 'For all x, if x is S, x is not P', and if these conditionals are rendered in the standard way as $-(\exists x)(Sx-Px)$ and $-(\exists x)(SxPx)$, then interpretations of A and E are true if their subject matters are empty, if there are no things in the world to which the subject terms apply. On the other hand, if 'Some S is P' and 'Some S is not P' are in turn understood as having the logical forms $(\exists x)(SxPx)$ and $(\exists x)(Sx-Px)$, the relations of subalternation between A and I and between E and O become invalid. 'All S is P' does not imply 'Some S is P' and 'No S is P' does not imply 'Some S is not P' unless there are things which are S. Nor can we simply reconstitute 'All S is P' as carrying with it the implicit assumption that some things are S without some loss; because satisfactory interpretations of 'All S is P' may be found that are consistent with there being no Ss. 'All offenders against this by-law are liable to prosecution' may be true although no instance of trespass ever occurs. And even if we were to re-read the universal form in the suggested way we could not by this step prevent the development of other distortions. In the traditional logic 'No S is P' is equivalent to 'No P is S' and therefore 'No mathematician is a person who has squared the circle' is equivalent to 'No person who has squared the circle is a mathematician'. But no one has squared the circle and therefore, on the requirement that subject terms must have members, the consequent must fail, and with it the convertibility of E. Preservation of the system would necessitate our saying that not only the subject terms but also the predicate terms are nonempty. Nor have we gone far enough along this path. If we allow negative terms, then from 'All S is P', with the aid of the traditional inferences of contraposition and conversion *per accidens*, we obtain 'Some non–S is non–P'. Must we then say that the traditional system can only be applied to nonempty terms whose negations are also nonempty?

If it is correct that the logic of 'Some S is P' is the same as that of 'Something is both S and P', then it is most reasonable to suppose, not that traditional logicians intended their logic to have just the limited applications forced upon it by this equivalence, but that they did not themselves fully understand the implications of their own forms and that had they understood these implications the traditional logic would not have developed in the way that it did. The alternative is possible, however, that the disabilities of the traditional logic arise not from the logical properties of the forms A, E, I, and O, properly understood, but from the modern treatment of them as interpretations of $-(\exists x)(Sx-Px)$, $-(\exists x)(SxPx)$, $(\exists x)(SxPx)$, and $(\exists x)(Sx-Px)$. If, for example, we read them as representations of, respectively, s[p(N)], s[–p(N)], s[p(M)], and s[–p(M)], then the question of the existence of things having the properties expressed by s, p, –s, and –p becomes irrelevant. The existence of such things is implied by none of these forms.

If this course is followed, the result is that all the relations of the square of opposition and all the immediate inferences and syllogisms commonly accepted in the traditional logic are theorems of the logic of general conditionals presented in secs. 1–4. Also, none of the invalid forms of syllogism is a theorem of OC. And this being so, it is possible to think that the logic of subject-predicate forms may be ultimately perceived as being internally consistent and as part of a larger system which includes the traditional forms as well as the more complex conditional and nonconditional forms of modern logic. Part of the interest of the logical correspondence between the traditional forms and their OC analogues lies, I think, in the fact that OC was not designed as a model of the traditional logic. The logical similarities of the two systems became apparent to me only after OC had been constructed.

The recommended treatment makes intelligible features of the traditional logic that have hitherto been thought to be straight mistakes. A characteristic objection to the traditional system is this. The I and O forms of the square of opposition are subcontraries, which is to say that they cannot both be false. It must be the case that either 'Some S is P' or 'Some S is not P' is true. Translated into standard predicate logic this statement amounts to the claim that 'Something is both S and P or something is both S and not P' is a logical truth. But this disjunction is equivalent to 'Something is S'. Thus if the disjunction were a logical thesis, 'it would be a theorem that something had [S], for *any* property [S], which is clearly absurd'.[4] Yet if the traditional logicians meant by 'Some S is P' that 'Something is S and P', and if the results are clearly absurd, it is surprising that they did not see that this is so. On the other hand, if 'Some S is P' is translated into OC as 'S is possibly P' in the sense of 'If something is S it cannot on that account be concluded that it is not P', I and O become, respectively, 'S is possibly P' and 'S is possibly not P'. Neither of them entails that something is possibly S, and

I and O become subcontraries. This is shown by the fact that 'S is necessarily P' (the denial of 'S is possibly not P') entails 'S is possibly P,' as we have just seen (Theorem 3(3)).

The question remains, of course, whether the system as so interpreted is 'right'—whether the traditional system, consistent though it may be, properly reflects the logic of subject–predicate statements. One feature of the suggested treatment of the traditional forms that readers may find hard to agree with is the assertion that 'Some S is P' does not imply that something is both S and P and that 'Some S is not P' does not imply that there is something which is both S and not P. The view is contrary to an interpretation of these expressions that has existed for hundreds of years.[5] Certainly the statement that some S is P does suggest the truth of 'Something is S and P', and in most cases when asserting 'Some S is P' or 'Some S is not P' we assume that there is at least one S that is P or that is not P—we have a confirming instance in mind. To see that such assumptions are not essential to the truth of 'Some S is P' or 'Some S is not P', it is helpful to consider these expressions as denials of the A and E forms. Let us grant that in asserting 'all S is P' we are sometimes saying no more than that there is nothing which is S and not P: that –(∃ x)(Sx–Px). And when –(∃x)(Sx–Px) truly expresses the form of the universal claim that is made, then (∃x)(Sx–Px) correctly expresses the form of its denial. But there are many contexts in which what is said by 'All S is P' is that if something is S, then one is justified (because, e.g., S is a cause of P) in concluding that it is P. And when this is the case, the denial of 'All S is P', although we may give expression to this denial with the words 'Some S is not P', does not say something about the existence of Ss that are not Ps, but rather denies the existence of the (causal or other) pattern that would justify one in concluding that a thing is P if it is S. Here the use of the word 'some' in 'Some S is not P' prompts us to think in terms of the instantiation of '. . . S and not P' (see p. 86) and to forget that 'Some S is not P' may properly be read as the denial of a relationship of dependency between S and P and not at all an assertion about the existence of things. It is this rendering of 'All S is P' and 'Some S is not P' as respectively the assertion and denial of a relationship, which, in making the logic of the syllogism coherent, is perhaps the one that is most appropriate.

Imagine a vender of patent medicines who mixes up a fresh batch of medicine for sale at a local fair. Before leaving for the fair grounds, he says to his wife, 'I don't think this new mixture is quite right—it doesn't have the usual taste and I don't propose to give it to anyone. Today I will simply take orders and deliver some better stuff later'. At the fair, before an audience, he declares, 'I say to you that anyone who uses this medicine', holding up the bottle containing the new mixture, 'will stay healthy'. Smith, a doctor, who hears this nonsense and is afraid that someone in the audience may take it seriously, speaks up, saying, 'That statement is surely false'. The vender now replies, 'In denying my statement you are claiming that there is someone who uses the medicine and will not stay healthy. But no one has ever used the medicine and no one ever will, because, the mixture being a bit off, I do not intend to sell it to anyone. Therefore, the statement that you are making is false and its denial, which is my statement, is true.' Smith would be warranted in responding to this by saying that the vender has claimed, not merely that there happens to be no person who uses the medicine and who will not stay healthy, but that the use of the medicine insures continued good health. It is this claim, says the doctor, that it is his intention to deny. If we take the doctor's side, it seems to me that we must allow that where the universal forms A and E are used to state connections between things, their denials, 'Some S is not P' and 'Some S is P', do not commit us to the existence of things that are S and not P or S and P. And because this approach to traditional forms—an approach that supposes that the A and E forms are appropriately used to state certain relations between things—brings coherence to the A, E, I, and O forms, I suggest that it is possibly the correct one.

Writers have also criticized the traditional process called 'inversion,' which licenses the inference from 'All S is P' to 'Some non–S is not P'—in our terms, from s[p(N)] to –(–s[p(N)]) or –s[–p(M)].[6] Why must it be the case that if all S is P, some non–S is not P? After all, if it were the case that everything is P, then all things S and non–S alike would be P.[7] But when 'All S is P' is taken to mean that a thing's being S warrants the conclusion that it is P, and when the problem is addressed from the point of view of OC, the inference becomes entirely natural. Because s[p(N)] means that the presence of s ensures the presence of p. And the only direct support that would give substance to this claim would consist of instances not only of s.p but also of –s.–p. And in the course of gathering this evidence we would have disproved –s[p(N)], and therefore proved –s[–p(M)]. Also, confirmation of the denial of –s[–p(M)] confirms the denial of s[p(N)]. The form, –s[–p(M)] is thus deducible from s[p(N)]:

3(9) $\dfrac{\text{s[p(N)]} \quad \text{entails} \quad \text{–s[–p(M)]}}{\text{Co is s[p(N)].–s[p(N)]}}$

Notice that neither component of Co is provable.

D(I)	I		DR
	s[p(N)]	⟶	–s[–p(M)]
Confirm s[p(N)] by –s1–p1	–s1–p1	TF implies	–s1–p1

D(II)	I		DR
	–s[p(N)]	⟶	s[–p(M)]
Confirm –s[p(N)] by s2–p2	s2–p2	TF implies	s2–p2

This question of inversion was not one that Aristotle had to consider directly. He doubted the equivalence of 'non–S' and 'not S' and his system did not admit of forms expressing negative terms. But there is some point in asking whether, had he accepted negative terms, he would have welcomed the inconsistency of 'All S is P' and 'All non–S is P'. He could have reasoned for the view that they are inconsistent by an argument similar to the one he used to show that if the statement A necessitates the statement B, then it cannot be true that the denial of A also necessitates B: "If 'A necessitates B' and 'Not A necessitates B' were both true, then not B would necessitate B".[8] Aristotle considered the consequent to be absurd and therefore rejected the possibility of the antecedent. But if one allows negative terms, and if we accept the consistency of 'All S is P' and 'All non–S is P', a similar result follows. From 'All S is P' we derive 'All non–P is not S' and from this and 'All non–S is P' we obtain 'All non–P is P'. It seems likely that Aristotle would have regarded this consequent as nonsense and that he would have concluded that 'All S is P' is incompatible with 'All non–S is P' and that the former thus entails 'Some non–S is not P'.

To establish the foregoing correspondences between the traditional system and OC does not show that OC offers the most acceptable rationalization of that system. But it is clear that when the language of 'all' and 'some' is replaced by the language of OC, the oddness of many of the transformations of the traditional system evaporates.

VIII

Let us observe the conjunction

(Q) b[c(N)].a[b(N)].a[–c(M)].

This is a normal inconsistency. (It is proved to be so in the Appendix.) No two components are inconsistent; the denial of any component is confirmed by the confirmation of the other two; and the proof of the only provable component, a[–c(M)], disproves the conjunction of b[c(N)] and a[b(N)]. The reader will notice that one of the entailments that can be derived from (Q) is the thesis of transivity, expressed as Theorem 3(6):

b[c(N)].a[b(N)]   entails   a[c(N)].

If we substitute in this formula for the predicates a, b, and c the symbol s, m, and p to represent the symbols for the traditional Subject, Middle, and Predicate terms, we have the first valid mood of the first traditional figure of the syllogism, *Barbara*, represented here as

m[p(N)].s[m(N)]   entails   s[p(N)].

By substituting in (Q) s for a, m for c, and p for b, we can obtain the syllogism *Baroco*, the fourth valid mood of the second figure,

p[m(N)] . s[–m(M)]   entails   s[–p(M)].

And by substituting in (Q) m for a, s for b, and p for c, we arrive at *Bocardo*, the fifth valid mood of the third figure,

m[–p(M)].m[s(N)]   entails   s[–p(M)].

C. Ladd-Franklin gave the name 'antilogism' to the type of conjunction that contains three components and is such that the denial of one component may be deduced from the assertion of the other two.[9] And she was able to show that fifteen of the twenty-four syllogisms commonly accepted as valid can be derived from five antilogisms. She assumed, as other logicians have assumed, that 'Some S is P' has the logic of 'Something is both S and P' and so was unable to deduce by these methods the remaining nine syllogisms in all of which particular conclusions are deduced from premises both of which are universal. These are *Darapti, Cesaro, Celaront, Felapton, Camestrop, Barbari, Bramantip, Fesapo,* and *Camenop*. The system OC, however, does not share this assumption about the logical effect of 'Some S is P', and in the Appendix I show how all twenty-four valid syllogisms can be elicited from eight normally inconsistent three-membered conjunctions.

Chapter 4

CONVENTIONS

I

I have taken the interpretations of the sentence letters of logical theory to be not sentences but statements. Sentences are transformed into statements when given reference, and as statements they become true or false. The sentence 'This is Tuesday' becomes a true statement when the pronoun refers to a Tuesday and a false one when it refers to a different day of the week. I have said too that to be a statement an expression must not only refer, but must also satisfy the application requirement—it must be possible to say or indicate what situations would confirm and what would disconfirm it. If this test is not met, the expression is not a statement.

In pursuit of a logic of factual conditionals, I have followed the practice of uncovering the forms of support and infirmation of conditionals by first finding their disconfirmations and constructing on these their confirmations. 'An east wind always brings rain' is falsified by fair weather on an east wind and therefore confirmed by rain following on an east wind and the absence of an east wind during fair weather. But there is a multitude of statements for which this approach is unsatisfactory. For instance, we could not accept as disconfirmation of 'All cats are animals' a cat that is not an animal because, as we say, a thing that is not an animal could not be a cat. A person who claimed to have located a cat that is not an animal would be said to have made a linguistic mistake, to have misunderstood the meaning, the accepted use of the word 'cat'. There *could* be no such thing as a cat that is not an animal. But if this is so, then 'All cats are animals' seems to contravene the application requirement, because if 'cat that is not an animal' is a predicate for which there could be no exemplification, 'All cats are animals' is not disconfirmable. And if it is not disconfirmable, it is not a statement and not something that can have entailment relations and therefore it cannot have a logic.

I will propose that assertions like 'all cats are animals' are properly regarded as formal logical truths, having the sense, 'The statement that a thing is A entails the statement that the thing is B', which are true by virtue of our acceptance of background statements that record the presence of linguistic commitments governing the use of expressions. These background statements are conditionals (though not factual conditionals) and they do have a logic. They are truths about the permitted or proscribed uses of expressions and I will call them 'conventions'. I do not mean by what I have said that 'This is a cat' and 'This is an animal', taken out of the context of the permitted uses of the expressions they contain, are such that 'This is a cat' (formally) entails 'This is an animal'. Employed in isolation, and without regard to the rules governing the use of 'cat' and 'animal', the latter is not formally deducible from the former. But our use of the predicative expressions 'is a cat' and 'is an animal' is controlled by conventions that limit the forms of confirmations and disconfirmations of the statements of which the predicates form parts. And, because in OC it is through the forms of the confirmations and disconfirmations of statements that we discover their entailment relations, once conventional considerations are taken into account, the influence of these conventions on the forms of confirmation and disconfirmation may produce entailment relations between statements that are not apparent when these statements are viewed in the light of their formal structure alone. But before showing how conventions influence the formal logical relations of factual statements, we must first establish that conventions do, in their own way, satisfy the application requirement and that they do have a logic. And before doing this we must find out what it is to be a convention.

Although the problem of the logic of conventions is one that falls within the scope of the general problem of 'meaning' in a language, it should not be thought that questions of meaning are all and only questions about conventions. Questions of meaning are questions about the purposes of language and the ways in which different parts of, and levels within language contribute to the achievement of these purposes. Contributions of the parts of a language to the creation of intelligible discourse are made in different ways and so one must not conclude that answers to questions about the nature and function of conventions are also answers to questions about, say, poetic meaning or about necessary or inescapable connections between expressions in a language. Conventional statements, as understood in this essay, assert commitments to use expressions in factual language in certain ways as a result decisions with regard to the use of those expressions by speakers of a language.

Why do we need this notion of the conventional statement in dealing with the relations between such expressions as 'cat' and 'animal', 'bachelor' and 'unmarried?' Could we not make do with the factual conditional and nonconditional statements of the type we have considered and also formal entailments that we could call logical truths? Why must we include, in our inventory of statements, these awkward assertions, which cannot be determined to be true or false by reference either to the objective world or to their formal structure? Suppose we were to adopt a language that consisted of singular terms, variables and one– and many–placed predicates ('–fx,' 'Gxy'), with complex statements being built up with the aid of truth functions and quantification and a suitable theory of conditionality. All true statements in this language would be either formal logical truths (those that are true and remain true under all reinterpretations of their components other than their logical particles) or synthetic statements about matters of fact. The result would be a language that is devoid of reference to conventional relations between expressions. Such an arrangement, if adequate, would certainly be convenient. Ordinary speech would be an interpretation, with modifications, of the propositional calculus and the first order predicate calculus, supplemented, if thought desirable, by a theory of conditionality. Presumably, however, it will be of interest to know how predicates in this language are to be applied. In order to make use of the claim, for example, that all Fs are Gs, we must be able to say what things are truly F and what truly G: before one can say whether or not all Fs are Gs, he must be in a position to justify his classification of things as Fs and Gs and to do this he must be able to say something about the application of the predicates 'F' and 'G', the method he uses to determine whether a thing is an F or a G. And here, one may say, is where the topic of the 'convention' introduces itself. A factual assertion can only be true or false if the expressions it contains are used correctly. A language, in the sense of a commonly adopted instrument of communication, cannot exist without agreement among speakers as to the uses of expressions. And statements as to the correct use of predicative expressions within a sentence are what I will call 'conventional' statements.

Our need to understand this type of statement becomes clear. Because how are we to know whether an assertion about the meaning of an expression is correct or not? One finds confirmation for 'This pen is blue' by examining the pen and finding out whether an appropriate observation is forthcoming. But how do we test the correctness of "One applies the predicate 'blue' to things like this [pointing]?" To begin with, not by simply examining the pen. The properties of the pen do not tell us whether the predicates we apply to it are properly used. It has sometimes been thought that the conventional meaning of a word is something in the mind and that a word is used correctly when its use is matched by a mental image. But this approach also seems unpromising. Suppose that whenever I describe something as being round I have an inward mental image of something square. This mental image does not change the meaning of 'round' nor does it make my use of 'round' if predicated of something round, incorrect. It is not an associated mental image that certifies that an expression has a certain appropriate application. Nor does an expression acquire its meaning from the mental or physical causes of the utterance of the expression, as I believe the following considerations will show.

Imagine that a psychologist has discovered the mental or physical causes of our verbal utterances. If this were to happen, and if statements about the sufficient conditions of our utterances could serve in the place of our statements about the meanings of words, our theory of discourse would be wonderfully simplified. We would be able to cast our statements about not only the connections between nonlinguistic events, but also about appropriate or prescribed relations between linguistic expressions, in terms of empirical law statements all testable in the same way and governed by the same logic. Statements of conventional meaning would be statements about verbal behavior, and the standards of application for them would be the same as those that hold for empirical law statements. But this avenue is not open to us. Even if psychological theory, or the physics of the brain, were sufficiently advanced to permit us to state the causal conditions of our utterances, we could not gain an understanding, from these statements alone, of which utterances were right and which wrong. They would not allow us to choose between Smith's statement, 'All bachelors are unmarried' and Brown's reply, 'No they are not'.

We might agree to let the majority win, and say that 'B' is the (or part of the) meaning of 'A' if all or almost all people are causally disposed to accede to the application of 'B' in a context whenever they are prepared to adopt the application of 'A' in that context. 'For naturalism the question whether two expressions are alike or unlike in meaning has no determinate answer, known or unknown, except insofar as the answer is settled in principle by people's speech dispositions, known or unknown.'[1] According to this proposal, we replace questions about the meaning of an expression with the sociological question about the speech dispositions of the majority. These dispositions are to be understood as caused tendencies to assent to the application of an expression when presented with a thing or with the application of another expression. 'A disposition is in my view simply a physical trait, a configuration or mechanism.' [2] This is a plan to treat those caused performances that consist of the universal or almost universal responses of a speech community to given stimuli as the ones representative of the meanings of the expressions in question.

There is a criticism of the dispositional analysis of meaning that is not quite to the point. This is that behavior is finite, and that no matter how many instances we amass of the acceptance by a person of B in a case of A, we cannot *show* that he will always accept the predicate B when he accepts A. On the next occasion, in another context, when different factors are present, or even when the circumstances seem entirely similar, he may accept A and reject B.[3] And if we cannot be sure what dispositions are true of one person, *a fortiori* we cannot be sure that his dispositions to use an expression in a certain way are the same as those of any other person. But this is a criticism to which all factual conditionals are equally vulnerable. And if the dispositional analysis is the only method we have for getting at the meanings of expressions, we must live with the method's limitations. Our response can only be Quine's: 'If by these standards there are indeterminate cases, so much the worse for the terminology of meaning and likeness of meaning'.[4]

But the disposition theory of meaning does have serious defects. For one thing, as I have said, it is not possible to reexpress the statement that under certain conditions a particular verbal response is appropriate or correct as a causal disposition statement. If I present you with an apple and ask you what it is and you say 'apple', your answer is correct. But the presentation of the apple and my question do not constitute a causally sufficient condition of the response. You *could* have replied that the thing was a pear or a space ship or you could have remained silent. We do not know how to state all the conditions that causally produce the response 'apple'; but we surely know that the causally sufficient condition, whatever it might be, would not be a reliable guide to the correct response. Causal conditions may inhibit the response 'apple' in the presence of an apple or they may result in the response 'apple' in the absence of one. We all know that this must be so because we all know that, when asked 'What is this?' we can say 'apple' when no apple is present, and refrain from saying 'apple' in the presence of one. Even W. V. Quine recognizes this problem to some degree when he says, '[W]e remain at the mercy of the speaker's veracity; we assume, when querying him that his assents and dissents are sincere.'[5] This is an admission that the prior state, if it is to be a causal one, must include factors that are in addition to the conditions that make a particular response correct. And indeed if sincerity must be part of the prior state, it is hard to see how the notion of meaning can ever be supplanted. Because to say whether a person's response is sincere or not involves making a judgment as to whether he knows the meaning of the expression.

Another way to see the inadequacy of the causal explanation of conventions is to notice that it nullifies our power to make distinctions that are crucial to any discussion of language. At one time almost all people were disposed to say that the sun travels around the earth. They were, one could say, caused to make such a response to questions about the sun, and on the causal theory, 'traveling around the earth' was part of the conventional meaning of 'sun.' And yet those from whom a different response was first elicited were not making a linguistic mistake about the meaning of the word 'sun'. Giordano Bruno was not persecuted for having made a mistake about the meaning of words.

Other characteristics of our linguistic behavior reinforce the view that rules for the use of expressions cannot be reduced to laws of psychological or neurological causation. Thus mistakes in following a rule of meaning do not disconfirm the rule but are breaches of the rule. But if following a rule were the same thing as being causally disposed to respond in a certain way, a breach of the rule would show that the claim that there is such a disposition is wrong. And if all people occasionally failed to respond in accordance with the disposition, this would have the result of showing that the expression did not have the meaning we thought it had; whereas we usually say, on the contrary, that the expression continues to have the same meaning but that we sometimes make mistakes in its use (because of the complexity of the rule, or for some other reason). This may be true even when only a linguistic minority uses an expression in a particular way. When a new term is introduced into a language, it is often the case that, for a time, a mere handful of people uses it correctly; the rest of us botch it. Thus statements about verbal behavior are not satisfactory replacements for statements of meaning. And although common verbal performances are aids to the determination of conventional meaning, they are so just to the extent that they evidence commitment to certain uses.

Related to the fact that nonfulfillment of a rule does not necessarily disconfirm it, is the additional fact that whereas causal disposition statements are descriptive, conventional statements are normative. If an expression is governed by rules, then we ought to use it in the prescribed way. It is because the establishment of a convention requires us to use a word in a certain way that failure to use it in that way does not itself displace the convention.

I observed at the beginning of this chapter that conventional statements do not respond to the type of confirmation and disconfirmation that serves for factual statements. We do not employ 'All bachelors are unmarried', for example, in such a way as to permit its disconfirmation by the observation of a married bachelor. And because I have adopted the view[6] that the logic of statements is governed by their direct support and infirmation, the convention will not, in OC, have a logic unless a form of direct evidence appropriate to it can be discovered. I have examined the possibility that psychological or behavioral statements, having the logic of factual generalization, might be found that could reasonably be said to take the place of conventional statements. Such statements, if they were unearthed, might discharge the need for conventional ones. The factual forms I have observed, however, do not have the requisite links with conventions. Conventions tell us what responses are required of us if we are to use a language correctly. And because conventions may change, because they are normative assertions that express rules that may be adopted or abandoned, they could not be causal. Causal statements about behavior are statements that tell us what will be done under complex circumstances, and it is not part of their function to say whether what will be done is correct or incorrect.

Instead, then, of trying to dispense with the convention, I shall treat it as a distinctive and perhaps indispensable type of statement and seek to establish its logic through the discovery of a form of evidence that can be received as its direct support and infirmation. And I will maintain that what justifies us in claiming that 'cat' refers to something that 'mammal' applies to or that 'aunt' means, by convention, 'parent's sister, or sister-in-law' as assertions about the appropriate use of predicates, is the acceptance by members of a language community of rules or regulations concerning the use of these expressions. It is because people accept these uses, and thus become mutually committed to them, that it is possible to say that such uses are binding on those people; and such assertions are what I have called 'conventions.'

Conventions are statements about prescribed uses of predicates and are supported by mutually accepted uses. But this type of support is not appropriate to all assertions of connection between predicates and the concepts to which they give expression. Philosophical issues about necessary connections between concepts have meaning, yet they are not to be resolved by statements that record commitments resulting from the adoption of regulations about the application of predicates. The problem, for instance, of the meaning of 'meaning' itself and whether this expression is eliminable in favor of behavioral ones, is not a problem about conventional meanings. Moreover, much of our discourse about ourselves and others cannot be elucidated in terms of conventional meanings alone. 'What did she mean by saying that her life has become a desert?' is a request for an explanation of the meaning of what has been said but it may legitimately be asked even though the questioner knows the conventional meanings of 'life' and 'desert'. The question in fact only becomes an interesting one if the conventional meanings of these expressions are already understood. Even sayings such as 'He was killingly kind' and 'He is a wise fool', which seem to be in breach of conventional regulations, have meaning. If this were not so, oxymorons would be senseless. Conventional meanings, however, have a certain priority. If predicative expressions did not have conventional meanings, their use in oxymorons and metaphors, to say nothing of assertions of philosophical import, really would be senseless.

Conventions derive from the acceptance of regulations; they are not mere regularities, habits. We may for the most part adopt rules without thinking, picking them up in the course of learning to speak, and having adopted them we follow most of them 'blindly'.[7] But rules can be reviewed, the appropriateness of following them reconsidered. It is a frequent characteristic of rules that they are followed unreflectively, but it is not a necessary characteristic. Old rules may be modified and new ones introduced. Many rules are superficial, adopted for the purposes of the moment. Shall we agree to say that a brick veneer house may be classified as a brick house or should 'brick house' refer to houses of solid brick construction only?

For the purposes of the interpretation of such and such a statute, should 'owner' of a building include a mortgagee in possession or only the beneficial owner? The rules governing the uses of other expressions—for instance, 'mammal' and 'insect'—are deeply entrenched and their change may involve far-reaching changes in the conventional meanings of other expressions. Neverthelsss, changes in well established rules do occur: the defining characteristic of water has not always been its molecular structure.

The rules from which conventions are derived are not necessary: we are not compelled by reason to adopt them and even after adopting them we may choose to alter them. Then how is one to justify conventional statements? Under what circumstances may a person say that it is correct that the speakers of a language are bound by a rule? We have seen that the justification is not to be found in the habitual nature of behavior. Regularity of performance is not equivalent to the adoption of a rule; regularity in itself does not permit us to say that an exceptional action would be in breach of a rule to which the actor is bound. Nor do one's intentions with regard to a practice determine its status as a rule in a language. An intention to adopt a rule is not sufficient to justify the claim that the rule is a rule of the language (e.g., if the intentional adoption is unilateral), and an intention to ignore a rule need not defeat it (if one is committed to it, along with others, as one of the standard rules of a common language).

I suggest that, assuming language to be a means of communication within a community of speakers, it is through our public acknowledgments, express and implied, that uses of words become committed uses—that as a result of these mutual acceptances, it becomes possible to say that certain uses are valid parts of a language. To the extent that our linguistic behavior involves express or implied undertakings to apply expressions in certain ways, certain forms of use become correct, others incorrect. That in coming to know a language we are concerned not merely with behavior of speakers but with committed usage; that it is not essential to rule following that it be blind, performed pursuant to immemorial custon; that usage is capable of being justified by our actions; and that the requirement that I now use an expression in a particular way is not to be derived from the past contents of my mind with respect to the future use of the expression—justification for all these propositions may be found in the following example.

Imagine two executives addressing themselves to the question whether to have a firm picnic. They decide that the picnic will be held if there is a consensus among the staff in favor of the picnic. 'What is to count as consensus?' Brown asks. Smith says that a consensus here should mean a favorable vote by not less than 90 percent of the staff and they agree on this. Having taken the vote, they find that only 80 percent of the staff are in favor of the picnic. Brown observes that on the basis of their previous decision about the meaning of 'consensus', the picnic will not be held, but Smith now says that they should hold the picnic because 80 percent approval counts as a consensus in favor of the outing. Brown objects that, for the purposes of the present issue, they had decided to treat 'consensus' as meaning not less than 90 percent. Smith responds that he is simply following, blindly, a customary rule as to the use of 'consensus,' and that when he agreed to use the expression differently the contents of his mind nevertheless determined that he would not use it in this new way. He says he is using the word 'consensus' to apply to 80 percent approval and that is that: there is no question of his being wrong in using it in this way. Brown replies that the previous agreement committed them both to the application of 'consensus' in a certain way and that this agreement justifies actual uses required by the rule and refutes inconsistent ones. It is Brown's argument that his own and Smith's behavior amounted to a mutual undertaking to use the word in a specific way, and that as a result of these undertakings, this use, in this tiny area of discourse, and during this tiny period of time, became the correct one.

If we accept Brown's position, as I think we must, this illustration serves to show that long usage is not necessary to the creation of rules of meaning. Nor is it necessary to rule formation that the speakers have particular intentions with regard to future usage. What is required is an understanding on the part of the speakers of what it is to become committed to a rule together with appropriate social behavior. From this we derive the meaning, within a fragment of discourse, of an expression; we learn how speakers ought to use the words of their language.

Of course, few rules are developed in the way Brown and Smith developed their *ad hoc* rule for the meaning of 'consensus.' We learn most of the rules we ever know by being told by others what these rules are and adopting them as our own and insisting that others use the expressions in question in the same way. Our commitments to rules *are*, for the most part, of long standing and our application of them is largely 'blind'. But automatic usage is an aid to discourse, perhaps a practical necessity; it is not a conceptually necessary feature of conventional meaning. The automatic, unthinking, use of rules makes it possible for us to attend to the world while talking about it without having to attend at the same time to the rules by virtue of which our talk makes sense.

Applications of expressions like 'A means B'—'consensus' means '90 percent,' "'divorced person' means 'person whose marriage has been dissolved' "—have a double aspect. In one respect they are encouragements, commands, instructions; they are acts of commitment and not as such statements about meanings, capable of being true or false. They are, in this sense, acts in pursuit of rules, bits of rule following and rule making. But to the extent that they are followed, agreed to, or not followed and rejected, it becomes possible to make statements with them. They become comments on the consequences, for the meanings of expressions, of the outstanding commitments with respect to their use of members of the speech community. It is in this second aspect that they may be taken to be true or false, as statements of conventional meaning. And statements of this type satisfy their application requirements or not on the basis of the commitments actually existing among speakers, commitments having their origin in the express or implied instructions, acceptances, acknowledgments, and insistences of speakers; and what conventional meaning statements say is not simply that certain regularities in verbal behavior occur but that these regularities are prescribed, that they are to be observed.

Mutual acceptance as revealed in our behavior does not, of course, establish or *prove* that the rule to be followed has a particular logical design. Let us imagine that our executives, instead of dissolving, through disagreement, their convention for applying 'consensus', make use of it in determining in future cases whether a consensus of the employees is for or against an office picnic. And then suppose that 90 percent agreement on an occasion is induced by some form of coercion, or the proposal is not properly understood by some of the employees who agree to it, or the vote is not with regard to an office picnic but some other form of entertainment. One cannot foretell how people will approach the applicability of 'consensus' in these circumstances. There will always be an element of doubt as to the effect of new conditions on our commitments. The fact of the accepted application of an expression in particular circumstances does not demonstrate that acceptance will persist when these circumstances occur in different contexts. Kripke has pointed out that the behavior of an individual does not prove that a rule is being followed. It is also true that mutual actions of commitment—the acceptance of certain applications, and the requirement that these applications be accepted by others—do not certify the correctness of applications of a rule in all contexts.

In reviewing conventions, and conceiving the possibility of their alteration, we become aware that there are restraints on the extent to which conventions can be modified and new conventions formed. One such limit is imposed by our need to make the world intelligible. We cannot, by the adoption of conventions, deprive the world of external objects having independent existence, or of other sentient beings. It is one thing, for example, to say that henceforth an institution may be classified as a public company even though not listed on the stock exchange; it is quite another to say that henceforth a thing may be counted as blue only if it is not extended. The latter position cannot be accepted without absurdity; its adoption would make the identification of colored things impossible. Another limitation on the formation of conventions is created by the structure of language itself, its syntax. We cannot adopt two conventions, for instance, if their joint application lands us in a contradiction. If this should occur, it is the conventions that must give way, not the underlying logical laws that issue in the contradiction. This may be illustrated by the following example. Let us assume that in our language (i) and (ii) are commonly accepted:

(i) All As are Bs, and
(ii) All Cs are not Bs.

As long as we find no things that are both A and C, we may accept (i) and (ii) as stating 'defining' or conventionally determined characteristics of As and Cs. But if we do find something, a K, which is both A and C, then either (i) or (ii) must be given up. No amount of agreement to retain both (i) and (ii) as conventions would make such retention correct or right. Suppose that (i) says that things (A) that swim in the ocean are (B) fishes and (ii) say that things (C) that lack gills are not fishes and that Ks are whales. If we retain both (i) and (ii) we are faced with the deductive conclusion that whales are fishes and not fishes. But we do not retain both (i) and (ii). It is the conventional arrangement that we give up, not the deductive consequence—if we did otherwise, our speech would become nonsense.

This illustration is taken from deductive logic; and it is the object of that discipline to seek out and identify those features statements must have, irrespective of their factual or conventional content, if they are to be coherent, noncontradictory. It is part of the business of philosophy too, I suggest, to establish those features of discourse that are necessary to the intelligibility of the language we speak—questions of the existence of material things, other minds, objective values, are questions, it seems to me, not of conventional usage but about what things we cannot avoid saying if our discourse is to make sense. The general question of the nature of philosophical arguments lies far beyond the scope of this essay. Something will be said about the nature of logical statements in chapter 6.

Conventional statements, the logic of which is developed in the following sections, are statements of the prescribed uses in a language of predicative expressions. Although they may be said to be statements of 'conventional meaning', they do not represent in themselves the only type of meaning statement and they are not attempts to say what linguistic meaning is, what it consists in. The conventional statement that if A is predicable of a thing, it is a rule that the application of B to that thing must be accepted, is not an entailment—its denial is not a contradiction. (As is shown in Theorem 4(6), however, when conventional arrangements are adopted as assumptions,they may contribute to the formation of entailments.) Nor, as I have noted, are conventions conceptual necessities. Conceptual necessities say that certain things must be the case, that they could not be otherwise. They are about the boundary markers of our capacity to understand the world ('External objects exist,' 'Past occurrences are not retrievable' are trite examples) or the inescapable structure of our language (as in, 'If A entails C and B entails not–C, then A and B are formally inconsistent'). What I have called 'conceptual necessities' are sometimes classified as entailments in works of logical theory, but they are not entailments in OC because their denials are not formal contradictions. Even claims about the formal structure of language such as 'If A entails C and B entails not–C, then A and B are formally inconsistent', although addressed to the issue of entailment, can be denied without contradiction and are thus not themselves entailments. As we shall see in chapter 6, substantive statements of logical theory are neither conventions nor formal entailments. Conventions show us what uses to accept in the light of choices that have been made; necessary statements attempt to show us areas of language from which the power of choice has been excluded. But although individual conventions are not necessities, the institution of the convention is an essential one in the sense that without its assistance, without some way of sorting out how predicates are to be applied, we would not have the privilege of being able to advance factual assertions capable of being true or false.

II

Of the many ways in which 'meaning' is employed in writings on language, it may assist us here to distinguish two. 'Meaning' is used in one sense when we say, or ask, what it is for a whole sentence to have meaning—what the characteristics are which would qualify the sentence as meaningful. In a second sense, we say, for example, that A means '. . .' when we wish to explain how A is to be used in the characterization and description of things. I use 'convention' or 'statement of conventional meaning' to stand for a limited class of meaning statements of the second type. Their role is to specify the way predicative expressions function in the production of factual sentences that are transformable into statements that are true or false. This should not be taken as implying that only factual sentences capable of serving as true or false statements have meaning in the first sense. There are many meaningful whole expressions that are not factual sentences and there are many ways of employing sentences other than as true or false statements. But I think it can be said that the conventions (expressions about meaning in the second sense), as studied here, concern a feature of language that is closely tied to a restricted or partial view of the function of sentences (which are the subjects of meaning statements in the first sense)—that function in which sentences serve as whole expressions that are transformable into true or false statements of fact.

We know that a sentence may be applied in a statement capable of being true or false when we understand what sort of evidence would support it when it is applied in a statement and also what sort of evidence would infirm it. And we know what a statement says when we understand the sentence it contains and know what the statement refers to. It is the application of a sentence to some aspect of the world that transforms it into a statement. This application need not take the form of an actual 'making' or propounding of a sentence. Actual assertion is not a necessary feature of statementhood. Counsel may ask a respondent at trial: 'Is it true that you committed adultery with the corespondent?' The respondent may then refuse to answer. But both respondent and counsel have considered, without asserting or denying, the statement that the respondent committed adultery with the corespondent. Thus an unasserted statement is nonetheless a statement.

We give the conventional meanings of individual words by showing how they contribute to the central work performed by statements. We show a person the meaning of the factual predicate 'red' by showing him how to identify red things and other things that are not red so that he can use 'red' in true statements. Because these acts of instruction, and complementary acts of acceptance, are common for a great many expressions throughout a language community, the members of the community may be said to speak the same language. We call the rules by which we establish the application of words for observables, like color words, 'ostensive' rules. When we have acquired some knowledge of these rules, we can also be taught 'testing' rules, or rules by the employment of which we are enabled to identify dispositional properties such as fragility, hardness, elasticity, buoyancy, and excitability. In learning how to use predicates that express dispositional properties we learn that statements that employ them can be determined to be true or false on the basis of a test—by putting a thing in water to see if it will float, or by pulling it to see if it stretches.

In fixing the meaning of predicates for observables and dispositional predicates, we also fix certain of the relations *between* predicates. If a patch of color is red, it cannot be blue. In learning how to use 'red' I learn that it is improperly applied to blue things. In this way, although countless conditionals remain factual, countless others become, implicitly, conditionals about the proper use of predicates. Whereas 'If it's heavy, Smith will not be able to lift it' is a factual conditional, 'If it's heavy it's light' exposes a mistake about meaning and 'If it's heavy it's weighty' is true by virtue of a feature of the rules governing the use of 'heavy' and 'weighty'. The truth expressed by 'If it's heavy it's weighty' is properly rendered by "If one applies 'heavy' to a thing, the conventions of the language are such that acceptance of the application of 'weighty' to that thing is prescribed". Assertions of this latter type are the statements I call 'derived rules' or 'conventions'; they convey the effect on language use of our commitments to rules. They are assertions that remind us of, or make claim to the existence of, rule governed relations between predicates.

Conventions are obvious; we employ them automatically and only occasionally require advice or assistance in their use. But they are not for this reason unimportant. They are essential devices with the aid of which we maintain order in discourse and insure that we say roughly the same things with the words we use. They are statements that are true by virtue of our commitments to employ expressions in certain ways. What they say is not that, for everyone, when A is used, B will be assented to, but that when A is applied, one ought to be prepared to accept the application of B. Thus their relations to their support and infirmation are different from the relations borne by factual statements to the events that confirm or infirm the latter. Conventions satisfy the application requirement not through confirmation by observed bits of verbal behavior, regarded simply as behavior, but through fulfillment by actions which are understood to consist in mutual acceptances of predicates in particular contexts.

III

Instructions, then, with respect to the use of words, and acceptances of these instructions, are not statements. But instructions and acceptances regarding particular uses give rise to commitments to accept, when certain expressions are applied, the use of certain other expressions. And assertions about these commitments, which I have called 'conventions', are statements to the effect that speakers of a language are bound by their commitments to particular uses. They are statements that are supported or infirmed by express or implied acceptances or rejections of speakers. But truly saying of a derived rule or convention that it is a convention of the language is not merely to make a statement of fact. When a person says, for example, that it is a rule that bachelors are unmarried or that whales are mammals, his statement, when it is a true statement, is not just a claim to the effect that speakers of the language use those expressions in this way. Part of what he is saying is that it would be a misuse of the rules to which we are committed to call a bachelor married or to identify as a whale something that is not a mammal. What a statement of this kind asserts—its objective state—is the existence of rules binding upon the speakers of the language. And the forms of the confirmations and disconfirmations of these statements—their observable aspects—are admissions by speakers concerning acceptable applications of expressions. These admissions may be implicit, and take the form of actual uses in ongoing speech, or they may take the form of express acceptances of actual or hypothetical applications. The notions of objective state and observable aspect are, I recognize, being brought to bear in a broad and inclusive sense: sufficiently broad in the case of 'objective state' to embrace normative requirements, and in the case of 'observable aspect' to include express and implied acts of acceptance of particular applications by participants in a discourse. This breadth is representative of features of explanations of conventional meaning found in natural speech. Our specifications of meaning do purport to say how predicates are to be used, without any thought of our being able to reduce these specifications to nonnormative form, and we appeal in support of these claims to those employments that would be required in actual or hypothetical cases.

I am maintaining that the statements to which we resort in passing judgment on the consequences of the adoption of rules are assertions to the effect that expressions of a certain sort deployed under certain conditions have conventional 'meaning'. And conventional statements are confirmed by the mutual acknowledgment of the correctness (on the basis of existing linguistic commitments to rules) of actual or hypothetical uses of the expressions the statements are about. If we accept a usage in our own speech, we may not reject it in the speech of others, and if we insist that a usage be observed by others, we are bound by the usage ourselves.

The statement that if one accepts the application of A to a thing it would be nonsense to deny B of it would be disconfirmed by a speaker giving an example, accepted by other participants in the discussion, which is not necessarily fulfilled in fact, of a use of A and not–B that might be true of something. Thus it might be possible to cite a hypothetical instance of a public company that would not be a company whose shares are listed on the stock exchange notwithstanding that the shares of all extant public companies were so listed, and thereby provide a counterinstance to the assertion that part of the meaning of 'public company' is 'company whose shares are listed on the stock exchange'. In urging this one might argue that 'public company' means 'company offering its securities to the public' and that the criteria for the application of 'offering securities to the public' do not require that the company be listed on the stock exchange. Generally the claim that an expression makes sense is confirmed by acceptable illustrations of its use and disconfirmed to the extent that we cannot find an acceptable application for it. Wrangles of this kind help us to achieve a better understanding of, as we often say, the 'meaning' of difficult expressions.

Rules of conventional meaning are not uniform throughout a language. In some circles the 'next–of–kin' of a person may extend to children of a spouse's brothers and sisters but in others only as far as the children of that person's own brothers and sisters. Superficial differences in the use of particular expressions do not induce us to say that a new language has been formed as long as these differences can be explained within the fund of conventions common to the speakers. Also, rules of language often change; and it is through illustrations of actual and potential uses in context that we keep in touch with such changes and are able to say, when changes occur, that what were once rules are rules no longer. The questions of what are acceptable applications are not settled in the same way in all regions of the same language. In some areas essential guidance is afforded by dictionaries, in law by decided cases, in science by lexicons unique to particular theories. But all these tools for the preservation of meanings themselves undergo gradual modifications to accommodate changes in the conventions they record. The central fact is that language is a cooperative social enterprise and questions of linguistic convention must be settled by agreement or not at all.

In the discussion that follows of the logic of conventions, it will be assumed that conventional statements are about predicates. They are about the permitted or prescribed uses of expressions which, or the negations of which, it is agreed are usable as predicates. Thus, to say of an expression D that it lacks conventional meaning in a context is not to imply that D may not be a predicate at all, but is rather to say that the denial of D is to be applied to all things of a given classification. If neither an expression nor its negation has conventional meaning, it is not a predicate. To say that where 'a widow' may be correctly applied to a person, it would not be meaningful to apply 'has never been married' to that person is not, on the usage I have adopted, to imply that 'has never been married' may not be a predicate. It is to assert that under these circumstances, application of 'has been married' to that person is prescribed. Employed in this way, 'conventional meaning' and 'prescribed' are related in a manner similar to that in which 'logically possible' and 'logically necessary' are related to modal logic. Thus to say that in given circumstances a formulation is not logically possible is not to imply that the formulation may not be a proper formulation at all but to assert that the negation of the formulation is logically necessary.

The risk that a conventional statement undergoes is that of being overturned by examples. The statement that an expression has conventional meaning is confirmed if its use can be illustrated by its incorporation as a predicate in a sentence that itself is capable of being used to say something true. It is disconfirmed if no use–in–context can be found. The statement that a predicate has no meaning, that the conventions governing its use insure that it could have no acceptable applications in the circumstances, is disconfirmed if one can discover *some* acceptable use for it and confirmed if its rejection is always correct. These confirmations and disconfirmations are statements that say that an actual or projected employment of an expression is or would produce a truth in context. This is why neither statements of conventional meaning nor their confirmations and disconfirmations are statements of fact, statements simply to the effect that a use has occurred; they are evaluations of uses. Their effectiveness as illustrations depends on the mutual acceptance of them by experienced speakers of the language.

In addition to assuming that the expressions of which meaningfulness, or lack of it, is asserted are proper predicates, the system that follows will also assume that the use of such expressions does not involve mistakes of category and that assertions or denials of conventional meaning are not about such mistakes. The assertion or denial of conventional meaning applies to expressions that are free of mixtures of type that make it absurd to say either that the expressions apply or that they do not apply. Thus we will never be concerned with such absurdities as that the number nine is asthmatic or that it is not asthmatic and I will never say of a predicate that it 'lacks conventional meaning' or that it 'lacks meaning' in an attempt to identify such absurdities. We will thus have to deal with the logic of a severely restricted concept of conventional meaning. But this simplified notion will afford us a device that will permit us to avoid mistakes that often occur when the logic of indicative statements is brought to bear on ordinary discourse.

I will construct conventions out of the modal operator 'C', which says of a predicative expression that it has conventional meaning. Where the letters 'D', 'E', 'F', and 'G' stand for predicates, that 'D' has conventional meaning will be represented by 'D(C)'. This tells us that under the working rules of the language, we can say that there are actual or possible things of which 'D' might be true—that the rules furnish us with a routine which, if it yielded a certain result, would assure us of the truth of 'D1' and, if it yielded another result, of its falsity, where '1', '2', '3', . . . stand for identified or hypothetical things or situations. 'D1' says not just that a usage occurs, but that the application of 'D' is accepted, a received usage. Because any accepted application will show that a predicate has meaning, the general form of the instantiation of D(C) is,

D1 v D2 v D3 v, . . .

where this form represents a finite collection of instances, and the general form of its counterinstantiation or rejection is,

–D1.–D2.–D3 . . .

To say that D has no conventional application, D(–C), is to say that its denial of everything is prescribed by the rules and the general form of the acceptance of *this* assertion is,

–D1.–D2.–D3 . . .

and that of its rejection,

D1 v D1 v D3 v . . .

Because the rejection of a statement is the acceptance of its denial, and its acceptance the rejection of its denial, –(D(C)) and D(–C) have the same form of acceptance and rejection. They are also related in the same way to their acceptances and rejections—their rejections establish them as false, whereas their acceptances, being finite conjunctions, leave open the possibility that an instance of rejection may be found. The form, –(D(C)) and D(–C) are therefore equivalent. To say that the use of D is prescribed—to be expressed by D(P)—is to say that the rules compel its application to everything and therefore the acceptance of this assertion has the form,

D1.D2.D3 . . .

and its rejection the form,

–D1 v –D2 v –D3 v . . .

D(–P) and –(D(P)) have the same form of acceptance,

–D1 v –D2 v –D3 v, . . .

and of rejection,

D1.D2.D3, . . .

and because they are related to these acceptances and rejections in the same way, they are equivalent.

As a result of these arrangements, D(C) is equivalent of –D(–P) and D(P) to –D(–C).

It is not difficult to see how the logic of nonconditional relations between predicative expressions emerges from these beginnings. The forms of acceptance and rejection of DE(C), D v E(C), DE(P), and D v E(P) for two instances are,

	Acceptance	*Rejection*
DE(C)	(D1.E1) v (D2.E2)	(–D1 v –E1)(–D2 v –E2)
D v E(C)	D1 v E1 v D2 v E2	(–D1.–E1)(–D2.–E2)
DE(P)	D1.E1.D2.E2.	–D1 v –E1 v –D2 v –E2
D v E(P)	(D1 v E1)(D2 v E2)	(–D1.–E1) v (–D2.–E2)

Let us attend here to a feature of these modal operators C and P, which has already been touched on. A significant use of D proves D(C); an actual or hypothetical application that would be accepted by the participants in a shared discourse, establishes that D has meaning and so D(C) is proved by the possible truth of a statement that instantiates it. But D(C) is not disproved by the form of its finite rejection, –D1.–D2, . . . A finite collection of cases illustrating the assertion that the expression 'payment combining principal and interest that is not a blended payment of principal and interest' is without meaning, does not assure us, beyond doubt, that the phrase lacks all significant application.[8] The reverse is the case for P. No collection of instantiations of D proves that there is nothing of which D may not be predicated. A single appropriate application of –D, however, shows that D(P) is false.

It is the conditional form of linguistic prescription, however, which receives the most frequent use, and whose logic presents us with the most difficult questions. We have seen that statements of conventional meaning are not statements of fact and so we cannot read "If 'is a bachelor' is predicated of a person, then we must accept the application of 'is unmarried' to that person", and "If 'it's a cat' is said of a thing, one must accept the application of 'it's an animal' to that thing" as nothing but alternative ways of saying 'If he's a bachelor, he's unmarried' and 'if it's a cat, it's an animal'. The form of the latter pair suggests what is not true of the former, that they are disconfirmed by married bachelors and nonanimalian cats.

Nor can we construct the logic of conventional conditionals on the base alone of truth functional connections between predicates. Suppose we were to try this and we said that D(P) entails E(P) if D TF implies E, where D and E take the same subject, and that 'If D is applicable to a thing, one must apply E to it' has the form –D v E(P). Then we would be faced with a puzzle of the sort that motivated the search for an alternative theory of factual conditionals. The meaningless application of a predicate to a thing would commit us to the application to the thing of any predicate. And the truth that the joint application of two predicates is prescribed would oblige us to say that the application of one of them is prescribed *if* the other is applied.

I propose, however, that the correct view is that the conventional conditional performs, for the logic of conventional statements, a function similar to that which the factual conditional performs for the language of fact. The factual conditional says that the world is such that the occurrence of one event ensures the occurrence of another event. The conventional conditional says that the system of language rules to which we are committed is such that the use of one expression ensures that it would be contrary to an accepted convention to withhold the application of another expression. The conventional conditional asserts our entitlement to say that one phrase must be treated as applicable when another phrase has been applied. Unlike factual conditionals, these statements about rules speak to normative, rather than factual connections between things. But, like factual conditionals, what they assert goes beyond what is observable in the evidence for them. The system formed by our commitments to ostensive and other rules is extremely complex. And no matter how certain we may feel of the dependency of one predicate upon another, this is not a relation that is open to inspection.

I shall take the conventional conditional (what I sometimes call simply the 'convention') as having the same logical form as the factual in that it is disconfirmed by a conjunction formed from its condition and denial of its result and confirmed by a conjunction formed from its risk factor and the denial of the remaining element in the disconfirmation. These confirmations and disconfirmations, however, are not composed of statements expressing nonlinguistic facts but of statements that assert that the actual or hypothetical applications of predicates are accepted in a language. And although the disconfirmation of a conventional conditional disproves it, its confirmations only support it and do not rule out the possibility of there being *some* actual or hypothetical application of its condition and denial of its conclusion that is appropriate. Conventional statements are not rules; they are comments on, fallible statements about rules.

It may be objected that in certain cases express commitments certify the conventions in question, so that there are situations in which actual and hypothetical uses do prove the truth of the conventions for which they provide evidence. If the executives just discussed both lived up to their commitments to treat 90 percent agreement as a consensus, would not this have proved a convention to the effect that 90 percent means a consensus? But what convention has been established here? (See pp. 117–87.) Would 90 percent agreement qualify as a consensus with regard to the purchase of an office coffee maker or a Christmas party? What if invitations to the party were accompanied by intimidation or unusual techniques of persuasion?

It is only as the rule continues to be applied in actual usage and hypothetical examples that we gather evidence of its structure, and even long usage does not prove the theory about its structure that is represented by the convention.

I apply to conventions the formal regulation, already adopted for (natural) necessity and possibility, to the effect that a schema containing one operator may be transformed into a schema containing the other by denying the whole, denying the conclusion and changing the operator. This means, for example, that the formulations without conditions, D(C), D(P), are equivalent respectively to –(–D(P)) and –(–D(C)). Therefore, " 'is blue' has conventional meaning" says the same thing as "It is not the case that the application of 'is not blue' to everything is prescribed" and " 'capable of being distinguished' is prescribed" says what is said by "It is not the case that to say of something that it 'is not capable of being distinguished' has conventional meaning". The prescriptive conditional D[E(P)] is equivalent to –(D[–E(C)]). The conditional D[E(C)], which says that if D is applied, one cannot on that account conclude that –E must be applied, is equivalent to –(D[–E(P)]).

D[E(P)] is not to be read, 'If D is applied, E is prescribed in all circumstances'; rather, the requirement to accept the application of E in a circumstance arises from the application of D in that circumstance. If one prefers, he may read D[E(P)] as 'It is prescribed that if D is applied, E is to be applied'. D[E(C)] is, in truth, the denial of a conditional relation and so the loose reading of it as 'If D is applied then the application of E has conventional meaning' is somewhat misleading in its suggestion that the meaningfulness of E arises from the application of D. The best and most accurate reading of D[E(C)] is the one just given, 'If D is applied, one cannot on that account conclude that the application of –E is prescribed'.

The confirmations and disconfirmations of –(D[E(P)]), –(D[E(P)], D[E(C)], and –(D[E(C)]) can now be seen to be as follows. (To simplify the presentation I leave out reference to the individuals to which the predicates apply.)

	Confirmations	*Disconfirmations*
D[E(P)]	DE, –D–E	D–E
–(D[E(P)])	D–E	DE, –D–E
D[E(C)]	DE	D–E, –DE
–(D[E(C)])	D–E, –DE	DE

IV

The definition of entailment with which we have become familiar applies to conventional statements, once allowance is made for the fact that the instances that exemplify or reject this type of statement are accepted linguistic uses and not the raw occurrences of nonnormative fact that formed the support for the statements of chapters 2 and 3. A conventional statement A entails a conventional statement B if each conjunction in the one or more termed disjunction formed from A.–B is inconsistent. And a conjunction is inconsistent if it contains a normal inconsistency. A conjunction is normally inconsistent if (i) the proof of each of its provable components TF implies the disproof of its other components, and (ii) the systematic confirmation of each implicans TF implies the confirmation in each alternative way of its corresponding DR.

Affirmative P statements, conditional and nonconditional, are subject to forcing. Affirmative C statements are not capable of being forced, but their instantiations will force the P statements with which they are conjoined.

The proof procedure is by now well understood, but it may be useful to illustrate it once more, by showing that if D has conventional meaning, and if the application of E is prescribed by the rules, given the application of D, it follows as a matter of formal entailment that E has conventional meaning.

<u>4(1)</u>	D(C).D[(E(P)] entails E(C)
	Co is D(C).D[E(P)].–E(P)

Only D(C) is provable and its proof disproves the rest of Co.

D(I)	I		DR
	D(C).D[E(P)]	⟶	E(C)
Confirm D(C) by D1	(D1)(–D1 v E1)	TF implies	E1

D(II)	I		DR
	D[E(P)].–E(P)	⟶	–D(P)
Confirm –E(P) by –E2	(–D2 v E2)(–E2)	TF implies	–D2

D(III)	I	DR	
	D(C).–E(P)	⟶	D[–E(C)]
Confirm D(C) by D3	D3.–E3	TF implies	D3.–E3

Other theorems, which will be obvious to the reader and that I state without proof, are the following:

4(2)	D[E(P)] entails –E[–D(P)]
4(3)	D(P).D[E(P)] entails E(P)
4(4)	D[E(P)].E[F(P)] entails D[F(P)]
4(5)	D[E(P)] entails D–E(–C)

4(2) represents the principle of contraposition for prescriptive conditionals and 4(4) that of transitivity. 4(3) shows that *modus ponens* also holds in this logic and 4(5) demonstrates that if on the application of D, E is prescribed, the joint application of D and –E is without conventional meaning.

It is not a thesis of the theory that the form 'If D is applied to a thing, then it is meaningful to predicate E of it' is itself transitive. Thus D[E(C)].E[F(C)] does not entail D[F(C)]. Consider the Co of this purported entailment:

D[E(C)].E[F(C)].D[–F(P)].

The exemplification of D[E(C)] on an occasion does not compel the exemplification of E[F(C)] and therefore instantiation of the denial of D[–F(P)] does not inescapably result. It is significant to say of a running thing that it is a deer and also of a deer that it is standing still; but it does not make sense to say of a running thing that it is standing still.

We saw in chapter 3 (pp. 100–101) that the principle of traditional logic that 'All f is g' and 'All g that is h is j' jointly entail 'all f that is h is j' does not hold in OC. The reader will notice that a corresponding principle fails for conventions.

D(E(P)].EF[G(P)]

does not entail

DF[G(P)].

That the position of OC is sound on this issue can be illustrated by any example in which F is, by convention, incompatible with D. If a thing is a sum of American dollars, it must, 'by definition', be a sum of money. And if a sum of money is composed of British currency, we must calculate it in pounds and pence. But we are not committed to saying that a sum of American dollars that is composed of British currency must be calculated in pounds and pence—'a sum of American dollars that is British currency' is, as things stand, an absurdity and the rules regarding its application say that it *has* no application.

V

The rules the force of which is asserted by conventional statements form part of the unexpressed environment of the language of fact. I stressed in chapter 1 that it is statements in context, not detached sentences, which are true or false and to which logical laws have their application. This environment molds the supporting and infirming evidence for factual statements by forcing that evidence to conform with assumed or 'received' facts and also with the rules that guide our search for the evidence. Statements are only right or wrong within that environment; if the assumed situation is not fulfilled—if the facts the statements assume are not facts, or if the parties to the discussion adhere to different language rules—then different statements are asserted and the intended statements are not made. Changes in context may change the structure of the confirmation and disconfirmation of a statement, in which case what is asserted by the statement will be altered.

In the review of the theory of confirmation in chapter 5, I shall give examples of the way in which neglect of the contextual surroundings of a statement has led to what seem to me to be unnecessary conceptual problems. But before proceeding to that discussion I shall set out here instances that show how factual discourse is shaped by derived rules and by assumptions of fact.

(1) Let us consider any member of the endlessly numerous class of statements that includes assertions of such triviality (e.g., 'if this is a tree, it's made of wood') that we rarely give them thought or expression. They are of interest primarily to Quine's 'unconversant persons' who are just beginning to explore a language they are attempting to learn. For all its obviousness, however, the fact that a statement such as this is true poses a problem. It cannot be a factual conditional because disconfirmation is precluded by the rules; and it is not a formal entailment because the joint assertion of antecedent and denial of consequent does not result in explicit contradiction. But its truth can be explained if we understand it as an assertion made within a context of derived rules that control the form of the evidential support and infirmation of its components. Let p stand for 'Smith is a bachelor', q for 'Smith is unmarried', D for the predicate 'is a bachelor,' and E for the predicate 'is unmarried.' Then by virtue of the conventional truth that D[E(P)], 'Smith is a bachelor' can be shown to entail 'Smith is unmarried'.

The conjunction of the proposed antecedent and denial of the consequent is p.–q. By the derived rule, it is true that D[E(P)], and so if D is applied, the acceptance of the application of E is prescribed. Now it is sufficient to show that p entails q if p.–q is inconsistent, and this will be the case if, when we rule out applications or denials of the predicates D and E that do not signify because contrary to prescription, the confirmation of p confirms q with at least equal strength and the confirmation of –q confirms –p in the same way. Let us exclude nonsignificant applications by enclosing them in the brackets '{' and '}'. Because D[E(P)] is true, D–E(–C) is also true, and this fact deprives p.–q of conventional meaning. Thus, if we express the confirmations of each of p, q, –p, and –q in terms of both p and q,

4(6) Given D[E(P)], p entails q

Co is p.–q

The proof of p proves q: pq v {p–q} TF implies q. Similarly, the proof of –q proves –p: {p–q} v –p–q TF implies –p.

	I		DR
D(I)	p	⟶	q
Confirm p	pq v {p–q}	TF implies	q

	I		DR
D(II)	–q	⟶	–p
Confirm –q	{p–q} v –p–q	TF implies	–p

Conventional statements are about acceptable and unacceptable uses of expressions. The language of fact presupposes that of conventional meaning in the sense that no statement of fact could be either true or false were there not accepted standards governing the use of the expressions it contains. I indicated at the beginning of this chapter (p. 110) my view that 'analytic' forms such as 'All cats are animals' should be treated as entailments. Theorem 4(6) shows the way in which this can be done. To the extent that 'All cats are animals' is a misleading way of saying "If the predicate 'This is a cat' is applied to a thing, the application of 'This is an animal' to that thing must also be accepted", then 'All cats are animals' is a conditional prescription about verbal expressions. But where 'is a cat' is a factual statement asserting that a thing has a characteristic, then because of an assumed prescriptive connection between the predicates employed, its relation to 'is an animal' is that of entailment. Of course, 'It's true that if a thing is a cat it's an animal' is grammatical in an everyday sense; my point is that this truth is not a factual conditional but is grounded in an entailment that assumes a convention.

(2) Rules influence the formation of confirmations and disconfirmations in different ways. We saw in chap. 2 that an item (p) cannot be a risk factor if it TF implies the truth or falsity of the remaining element in the disconfirmation of a conditional (say –p). Fulfillment of a risk factor must create a contingency, a risk, and it cannot do this if the truth or falsity of the remaining element follows from it as a matter of formal deduction. This is why, in the discussion of Stalnaker's theory in chapter 2,[9] I rejected the claim that 'If p then q', in the sense of a contingent conditional, is deducible from 'p entails q'. If p does entail q, then 'if p then q' cannot be a contingent conditional at all and therefore cannot be understood as making a contingent assertion and so cannot be either true or false. The same holds if the truth or falsity is prescribed for the remaining element by derived rules. It is because 'He is a bachelor', assuming the governing rules, entails 'He is unmarried' that 'If he's a bachelor, he's unmarried' is not a factual conditional.

(3) In the discussion of 'disposition' conditionals in chapter 2 (pp. 55–56) I noted that a feature of the disposition conditional that distinguishes it from the ordinary one is that one can deduce, from the denial of a disposition conditional, for instance, 'If you place this in water, it will float (in the water)', the truth of 'If this is placed in water it will not float (in the water)'. This deduction is not true of the ordinary conditional because –(p[q]) is fully confirmed by p–q, whereas the full confirmation of p[–q] requires that it be confirmed in one phase by p–q and in another by –pq. Also –(p[q]), but not p[–q], is proved by its confirmation and so even if the confirmation of –(p[q]) and p[–q] were the same, the confirmation of the antecedent would not confirm the consequent with the same force. The peculiarity of the disposition conditional may now be explained. In the case of the sort of disposition conditional referred to in that discussion, the sense of what is said in the condition is partly repeated in the conclusion—we place the thing in the water to see if it floats (in the water). This being so, an assertion that seeks to state the nonfulfillment of the condition p and fulfillment of the conclusion clause q will lack conventional meaning—a thing cannot be not in the water and floating in it. Then p[–q], not being confirmable by –pq, takes from p–q all the confirmation it is capable of receiving and is

thus fully confirmed by what confirms –(p[q]). The form, p[–q] is also proved by p–q, if one allows, as I did in the examination of the subject in chapter 2, that confirmation of a disposition conditional proves it.

But these thoughts answer only to half of the question. The Co of '–(p[q]) entails p[–q]' is –(p[q]).–(p[–q]). I have shown that the confirmation of –(p[q]) may be said fully to confirm p[–q], but not that the confirmation of –(p[–q]) fully confirms p[q]. The form, –(p[–q]) is proved by the fulfillment, during the occasion, by a complex event of the form pq, whereas p[q] must be confirmed by –p–q(i) during a phase and by pq(ii) during the other phase. The answer to this problem may be as follows. The subject of analysis is the sort of disposition conditional—'If you pull this it stretches,' 'If you put this in water, it dissolves'—which says that on the fulfillment of a test, a result will be revealed. And this conditional assumes, I suggest, that at some phase of the occasion to which it refers, the test has not been fulfilled. And because the nonfulfillment of the condition and fulfillment of the conclusion (–pq) lacks, as we have seen, conventional meaning, the situation during the phase at which the condition has not been fulfilled must have the form –p–q. Thus the assumed context of p[q] already includes its confirmation by –p–q(i). These arguments do not perhaps fully explain this distinction between ordinary and disposition conditionals, but they do indicate that it is important that we attend to received factual and conventional aspects of the assumed context of factual statements when trying to come to terms with their logical consequences.

(4) The disposition conditional may be regarded as a special case of the 'suppositive' conditional the general characteristic of which is that the condition and conclusion clauses are so related that the fulfillment of the conclusion or of its negation conventionally involves the fulfillment of the condition (see p. xiv, pp. 44–45). 'If it rained then it would not rain heavily' implies 'It rained'. And Frank Jackson[10] has used it in an argument against the validity of contraposition: from 'If it rained it would not rain heavily' one may not deduce 'If it rained heavily it would not rain'. But the example seems to me, rather, to illustrate the truth that logical laws cannot survive applications that ignore the conventional relations between the predicates of the sentences they employ. *Modus tollens* for conditionals, even material conditionals, would collapse into absurdity under the weight of this type of interpretation. A logic of conventional relations of the kind presented in this chapter allows us to

show how conventional relations between predicates affects the logical relations of factual statements by identifying the influence of conventions on the forms of confirmation and disconfirmation.

(5) The context of a statement, as I have suggested, contains besides rules, factual situations to be taken as given by the parties to the discussion. Some factual assertions expressly indicate a fact that is to be assumed as being in place for the purposes of assessing the information being considered. Counterfactuals, for example, show by the manner of their construction that we are to assume that the antecedent is not fulfillable. 'If Hitler had invaded England in 1940, Germany would have won the war' is to be understood on the assumption that Hitler did not invade England in 1940. It therefore cannot be disconfirmed and can only be confirmed in one alternative way, by the noninvasion of England in 1940 and by Germany's failure to win the war. Singular counterfactuals are thus partially disabled as empirical claims and cannot command strong allegiance unless supported by the indirect evidence of well confirmed universal conditionals.

I suggested in chapter 2 that we may think of the logic of the ordinary conditional as being equivalent to that of the subjunctive conditional, 'If p were to be the case then q would be so'.[11] What matters is the objective of the statement. If it is the point of the indicative conditional and its corresponding subjunctive form to assert a relation between condition and conclusion that ensures that the conclusion is fulfilled if the condition is fulfilled, then the two statements have the same logic. They will differ in their prospective confirmations, however, if the subjunctive form is assumed also to be counterfactual. In that case one of the potential confirmation forms of the subjunctive conditional will be precluded. Ernest Adams's theory[12] that the indicative and subjunctive conditionals have different logics, however, turns on a different issue. He points out that we are inclined to accept the first of the following conditionals but not the second.

(1) If Oswald did not kill Kennedy, then someone else did.
(2) If Oswald had not killed Kennedy, then someone else would have.

Let us observe, however, that the contexts of these statements are quite different. The second, being counterfactual in form, asks us to consider (contrary to the assumed facts) a situation in which Kennedy has not been assassinated and states that in that factual context, if Oswald does not kill Kennedy then someone else will. If we reject the statement, we do so because we lack sufficient reason to believe that the social realities of that occasion entitle us to suppose that the events in question are so related. The source of the plausibility of (1), however, is the assumption that *someone* killed Kennedy. Given this assumption, the statement that Oswald did not kill Kennedy *entails* that someone else did. (1) if true, is true as a logical deduction, not as a conditional. If the assumption is disallowed and if we are to consider (1) as asserting simply a conditional relation, we have no more confidence in its truth than we have in that of (2). Subjunctive and indicative ordinary conditionals assert the same relation; the contrary to fact subjunctive conditional indicates that we are to ignore certain facts, known to be true, in making a judgment as to the presence or absence of that relation in a situation now in the past.

Chapter 5

CONFIRMATION

I

In previous chapters the concept of confirmation has served as the medium through which entailment relations between statements are established. Let us now approach the question of confirmation from the point of view of the justifying effect a singular confirming statement has on the general statement it confirms. Addressed from this standpoint the question of confirmation divides into two parts, the first being the question of the form a singular statement must take if it is to provide an increment to the direct empirical ground or justification for the assertion of the general one, and the second being that of the evidential strength of a collection of singular statements, the degree to which they justify the adoption of the statement they support.

I will be directly concerned only with the first of these questions. And the reader will by now be in a position to anticipate the direction my answer will take. My central proposal has been that a fact, to qualify as a form of confirmation of a statement, must be capable of being read as a response to a risk of disconfirmation. And where this response may take alternative forms, we do not regard the statement as having been confirmed in a way that to any degree justifies confidence in its truth, unless the statement has at least been confirmed in each alternative way. This theory has played a principal role in the development of the concepts of conditionality and entailment. What I mainly wish to do in this chapter is to attempt to explain where I think standard theories of confirmation have gone astray. I will argue that previous accounts have erred in ignoring not only the nature of conditionality but also the influence that derived rules or conventions have on the application of logical principles to natural language. It is for this reason that I have postponed the discussion of confirmation theory until after the discussion of conventions in chapter 4.

As long as we restrict ourselves to nonconditional statements, the issue of confirmation does not raise serious difficulties for the traditional predicate calculus. We find the form of the singular confirmation of a quantified monadic predicate schema by removing the quantification sign and replacing the variable by a symbol for the name of an individual. The quantified form is disconfirmed by the denial of its confirmation. In the language of the present theory, the logical form of a confirming instance of a nonconditional generality is the same as that of the subject matter of the generality and its disconfirmation has the form of the denial of the subject matter. If the generality is an existential one, the effect of any confirmation is to prove it; a single confirmation logically implies the long disjunction that is the form taken by the confirmation of a particular statement. But the effect of any disconfirming collection is not to disprove it but rather to make its truth less likely; no finite collection of conjunctions of singular forms logically implies the conjunction of indefinite length that must be true if the particular statement is false. By the same reasoning, a finite collection of confirming conjunctions supports but does not prove the nonconditional universal it confirms; but a single disconfirming instance disproves the universal. (The matter becomes more complicated when existential and universal quantifiers are mixed in polyadic contexts. 'Everything x is such that there is something y, which bears R to x' can neither be proved nor disproved.)

This account of confirmation is essentially the same as the one given by Hempel in his famous "Studies in the Logic of Confirmation,"[1] where direct confirmation is defined in these terms: 'An observation report B directly confirms a hypothesis H if B entails the development of H for the class of objects that are mentioned in B'. And he says that the 'development' of H for a finite class of individuals C states what H would assert if 'there existed exclusively those objects that are elements of C'. Take, for example, the development of the hypothesis $H_1 \equiv [(x)(P(x) \vee Q(x)]$ which says that every object has the property P or the property Q. For the class $\{a,b\}$ this is $[P(a) \vee Q(a)]$. $[P(b) \vee Q(b)]$, that is, *a* has the property P or the property Q, and *b* has the property P or the property Q. Of course if H_1 is an open universal, it is not proved by confirmation by a finite number of items. And we may argue that confirmation by a finite collection of instances affords us no ground for supposing that H_1 will be supported by future instances. Empirical support in this form, however, at least allows us to say that H_1 has been confirmed 'so far' by items that are relevant to its truth.

But an important problem is posed for this method of defining confirmation when, as in standard systems, conditionality is rendered as a disjunction. Because then 'If anything is P it is Q' becomes (x)[–P(x) v Q(x)] and the development of this for the class {*a,b*} is [–P(a) v Q(a)]. [–P(b) v Q(b)]. But this in turn is implied by [–P(a).–P(b)]. Thus every nonfulfillment of the antecedent of a universal conditional confirms it. And so, whereas in ordinary discourse we would say that if John never plays bridge we have no direct evidence for (X) 'Whenever John plays bridge, he wins', by the standard account of confirmation, if John lives his entire life without ever playing bridge, we have a substantial body of evidence which confirms (X). This is a consequence of the standard approach to confirmation that not many are prepared to accept. The problem here is not that our 'confirming' items are not sufficient to establish the truth of the generality. The trouble is that the items themselves are irrelevant; they do not even justify the claim that the generality has been confirmed so far, or up to the present. The difficulty has received the name 'The Paradox of the Ravens'—every nonraven, every leaf or handkerchief, confirms the hypothesis that all ravens are black. This is perhaps not a paradox in the ordinary sense; it does not mark an inconsistency in the Hempelian theory of confirmation. What it shows is that the Hempelian theory, with respect to this question, is not one of which use is made in actual discourse. The example of the ravens illustrates the fact that acceptance of Hempel's theory would compel us to adopt practices that we do not follow, and that we would rebel against following, in our ordinary and scientific reasonings about matters of fact. In this chapter I will review this and other problems encountered by the Hempelian theory and some of its variants and also try to show why these problems do not affect the theory of confirmation that in chapters 1 to 3 grew out of the attempt to represent the logical relations of factual conditionals.

If we do not accept nonfulfillments of the antecedent as instances that confirm a conditional hypothesis, what instances do we accept? The naive reply to the question: "What confirms 'All beavers have tails' "? is invariably 'Beavers with tails'. And this has been the reply of at least some logicians. It was the view of Nicod, for example, that 'If A then B' can only receive direct confirmation from the presence of B in a case of A. "Consider the formula or the law: A *entails* B. How can a particular proposition, or more briefly, a fact, affect its probability? If this fact consists of the presence of B in a case of A, it is favorable to the law '*A entails* B'; on the contrary, if it consists of the absence of B in a case of A, it is unfavorable to the law. It is conceivable that we have here the only two direct modes in which a fact can influence the probability of a law."[2]

We must take the word 'entails' in this quotation as having the same meaning as, 'implies' traditionally has when it is used to express an empirical, rather than logical, connection between properties. In this sense it is simply the generalization of a material implication and 'A entails B' may be symbolized by (x)(Ax ⊃ Bx).

But (x) (Ax ⊃ Bx) and (x) (–Bx ⊃ –Ax) are logically equivalent. And an objection to Nicod's theory is that if he is right, a case of Ax.Bx confirms (x) (Ax ⊃ Bx) but not its equivalent (x)(–Bx ⊃ –Ax). The approach then fails to satisfy what has been called the 'Equivalence Criterion' of confirmation: that what counts as confirmation of an empirical generalization X must also count as confirmation of any generalization that is logically equivalent to X.

Nelson Goodman has also recommended a solution that implies rejection of the equivalence criterion. On this view, no instance confirms 'All As are Bs' if it also satisfies 'No As are Bs'. Because these are equivalent, in classical logic, to 'Everything is either not–A or B' and 'Everything is either not–A or not–B', respectively, the only thing that will confirm the former without also satisfying the latter is an instance of A and B. In this way, confirmation of 'All As are Bs' is reduced to instances of As and Bs and confirmation of 'All non–As are non–Bs' is reduced to things that are both not–A and not–B.[3]

W. V. Quine has proposed the different view that satisfaction only of predicates that are 'projectible' can serve as confirmations of generalities. He says that 'raven' is projectible and that therefore 'All ravens are black' is supported by instances of black ravens, but that 'nonblack' is not projectible and as a result 'All nonblack things are nonravens' is not supported by nonblack nonravens. It can only be confirmed indirectly by what confirms its equivalent, 'All ravens are black'.[4]

It is true of each of these approaches, however, that although they succeed in excluding, from the class of confirming instances of 'All As are Bs', certain cases that seem irrelevant to its truth, these cases continue to confirm generalities which entail 'All As are Bs', in the reading of this form that gives it the same logical structure as (x)(ax ⊃ Bx), and the undesirable cases, therefore, will provide as strong a ground for the acceptance of 'All As are Bs' as they do for the entailing statements. Thus instances of non–A do confirm 'Nothing is A' and this generality entails (x)(Ax ⊃ Bx). And so instances of non-A afford a ground for acceptance of (x) (Ax ⊃ Bx) which is no weaker than our justification for the acceptance of 'Nothing is A'.

But I have another objection to these solutions that I am sure the reader will have foreseen. It is this. Nicod, Goodman, and Quine have understood the importance to the confirmation of 'All As are Bs' of the absence of instances of Ax.–Bx and also of the presence of instances of Ax.Bx. But what their views do not apprehend is that the joint fulfillment of condition and conclusion of the universal conditional does not *in itself* help to justify the assertion that the fulfillment of A results in the fulfillment of B. The Humean notion that the 'constant conjunction' of As and Bs is what confirms 'All As are Bs' is convincing to the extent that we think of As and Bs as properties that things do not always have. If everything one could find were both A and B, this would not encourage us to say that a thing's being A is responsible for its being B. It is against the background information that all known non–Bs are non–A that the Humean maxim is plausible. From the perspective of OC, the significance, for the confirmation of a universal conditional, of the joint fulfillment of condition and conclusion derives from the general principle that X qualifies as a confirmation form of a universal conditional Y if it can be interpreted as a positive response to the exposure of Y to the risk of disconfirmation. It is when we find that the fulfillment of the risk factors of a conditional are always accompanied by risk avoiding properties—when As are consistently Bs and non–Bs consistently non–As—that we obtain substantive confirmation of 'All As are Bs.' Ax.Bx is the form of a confirmation-part of 'All As are Bs' and so is –Ax.–Bx. But the fulfillment A1.B1 does not supply a portion and –A2.–B2 another portion of the confirmation of the universal (see chapter 1, p. 15). It is the combined effect of these fulfillments that produces substantive confirmation. And so 'All ravens are black' and 'All nonblack birds are nonravens', being logically equivalent, also have the same confirmations and the same disconfirmation, neither is entailed by 'Nothing is a raven' and neither is 'confirmed by', in the sense of 'receives substantive support from', the discovery that some nonraven is nonblack, unless this is accompanied by the additional discovery of black ravens. It is because instances of A and of non–B nearly satisfy, come within one property of satisfying, the disconfirmation of the universal, that instances of As conjoined with Bs and of non–Bs conjoined with non–As do support the universal and that instances of non–As conjoined with Bs do not. Also, we should not overestimate the significance of the form of confirmation that consists in the satisfaction of the condition and the result. It is when we have accumulated instances of both types of confirmation that we begin to justify the claim that the occurrence of the condition is relevant to the occurrence of the result. This is so, at least, when part of what one is asserting by, for example, 'All ravens are black' is that a thing's being a raven is sufficient to justify one in saying that it is black. If all birds are black, so that we could find no cases of nonblack birds that were not ravens, we could not suppose, no matter how many black ravens we found, that in particular the ravenhood of a bird had anything to do with its being black. This point is well illustrated by situations in which a universal conditional can be tested by human intervention. The proper testing of 'When this chemical is used, it kills vegetation' involves the application of the chemical to healthy vegetation, and so in contexts in which the chemical has not been used and the vegetation has not been killed. The application of the chemical to vegetation already dead would not be regarded as constituting a significant test.

(Someone may object that if everything is B and some individual (say 1) is A, *then* it must be B, and so confirmation of 'Everything is B' must support a conditional relation between A and B. But this argument overlooks the fact that 'Everything is B and 1 is A' *entails* '1 is B' because 'Everything is B' alone entails '1 is B'. The *then* of 'Everything is B and 1 is A, *then* 1 is B' separates the antecedent and consequent of a formal deduction, not the condition and conclusion of a contingent, conditional relation between A and B. It has been with a view to escaping confusions of this type that I have avoided the use of the 'If . . . then . . .' formulation in stating entailments. We should also remember that if everything is B, then 'All A is B' is true in a material sense: 'Everything is B' entails 'Everything is –A or B.')

Quine has said that green leaves, which are nonblack and nonravens, are not evidence in support of 'All ravens are black'.[5] In OC a nonblack nonraven, and therefore a green leaf, fulfills one of the confirmation forms of 'All ravens are black'. But there is an oddity about the example that relates I think to the way in which we seek out situations that test empirical hypotheses, the way in which additional evidence is accumulated. If we are to place the claim that all ravens are black in jeopardy we must seek out situations in which ravens that are nonblack things *may* be discovered. We do not look for ravens among things already established to be sparrows, or armchairs; and we do not look for nonblack items among things that have characteristics that preclude them, by convention, from being ravens, because production of such things cannot imperil the status of 'All ravens are black'. A fact expressible by a statement that entails the denial of the remainder of the disconfirmation cannot (as I have argued in chapter 2) qualify as a risk factor. And 'This is a leaf' does, for the reasons expressed in chapter 4,[6] entail 'This is not a raven'. In trying to find tests for the hypothesis we must at least look among feathered vertebrates, among things having the general properties the detailed examination of which will lead us to the judgment that they are or are not ravens. Tests conducted among things known not to be birds are redundant. As an item of confirmation the green leaf is odd not because it is a nonblack nonraven but because it is superfluous. It represents a test the conclusion of which has been determined before it begins.

It is a truth of classical theory that, because (I) (x)(Ax ⊃ Bx) entails (II) (x)(Ax ⊃ Bx v Cx v . . .) and (III) (x)(Ax.Cx . . . ⊃ Bx), then no matter what theory of confirmation is adopted, the acceptance of (I) commits us to acceptance of (II) and (III). It is not possible to respond to these deductions by saying that from the facts, whatever they may be, from which we actually project the hypothesis (x)(Ax ⊃ Bx), we do not also project (x)(Ax ⊃ Bx v Cx v . . .) or (x)(Ax.Cx . . . ⊃ Bx). Because by the common standard of deduction, the latter forms are the consequences of the former so that to the extent that we permit any evidence to commit us to the former, commitment to the latter cannot be avoided. Nor is there anything to regret in these deductions as long as '. . . ⊃ . . .' is treated as a sign of material conditionality, as the logical equivalent of '–. . . v . . . .' But the case is different if '. . . ⊃ . . .' is interpreted by the language of sufficient and necessary conditions: if, for example, the part of the formulation following '⊃' is adopted as a necessary condition of the part that precedes it, so that the preceding part stands to the following part as an empirically sufficient to a necessary condition. Material conditionality will not do justice to this relation. Because it would mean that 'B is a necessary condition of A' would entail 'B or C forms a necessary condition of A' and, by transitivity of entailment, would also entail 'A and not –B form a sufficient condition of C', which I suggest is clearly a distortion. Also, if 'A is a sufficient condition of B' is to entail 'A and C form a sufficient condition of B', then by similar reasoning it will also entail 'A and not–B form a sufficient condition of not-C'.

Writers who have considered these results have thought them inescapable if not wholly desirable. G. H. von Wright, for instance, has said that 'It is clear that if a property P is a necessary condition of A, then the disjunction of P and *any* property is also a necessary condition of A. For, that P is a necessary condition of A means that in all positive instances of A the property P is present, and of all instances where P is present it is trivially true that P *or* P^1 is present' and again 'If a property P is a sufficient condition of A, then the conjunction of P and *any* property is also a sufficient condition of A. For, that P is a sufficient condition of A means that in all positive instances where P *and* P^1 are present it is trivially true that they are positive instances of P'.[7]

These consequences are inescapable if we regard statements of empirically sufficient and necessary conditions as interpretations of material conditionality. But they are not conclusions of the present theory of ordinary conditionals that emerges from the marriage of the concepts of confirmation and entailment. In this theory, a body of true singular statements supports a universal conditional if,

(i) it contains statements that confirm the conditional in each alternative way, and
(ii) it contains no statement that disconfirms the conditional.

Under this system, the model form A[B(N)] is not confirmed by either of the forms –A . . . or B . . . but only by A . . . B . . . and –A . . . –B . . . . And although each of these qualifies as a type of support, neither contributes significantly to the positive ground for acceptance of A[B(N)] unless it is part of a conjunction of statements containing instances of the alternative types of confirmation and lacking instances of disconfirmation. In the second place, the theory overcomes the problem that the equivalence criterion posed for the Nicodian definition because each instance of A . . . B . . . and –A . . . –B . . . is a confirmation form of both A[B(N)] and –B[–A(N)] and neither of the latter forms is confirmed by forms other than A . . . B . . . and –A . . . –B . . . .

Thirdly, the confirmation of A[B(N)] does not confirm either AC[B(N)] or A[B v C(N)]. Neither A . . . B . . . nor –A . . . –B . . . TF implies either A . . . B . . . C . . . or A . . . –C . . . –B . . . or –A . . . C . . . –B . . ., the confirmations of AC[B(N)]. Nor do they TF imply the confirmations of A[B v C(N)], A . . . –B . . . C . . ., A . . . B . . . –C . . . and –A . . . –B . . . –C . . . . Also, we have seen that A[B(N)] does not entail either AC[B(N)] or A[B v C(N)].

Where *X* and *Y* are universal conditionals, the entailment of *Y* by *X* insures that the confirmation of *X* TF implies the confirmation of *Y*. But the converse is not true. If *X* entails *Y*, not only must the confirmation of *X* confirm *Y*, but the disconfirmation of Y must disconfirm *X*; also the proof, if any, of *X* must prove *Y* and the disproof, if any, of *Y* must disprove *X*.

II

Great attention has been paid in writings on the concept of confirmation to a puzzle which, it is said, exposes a weakness in our ability to construct acceptable confirmations of universal conditionals. The puzzle in fact disappears when we recognize that the weakness to which it refers does not attach to the procedures by which we actually go about collecting evidence for or against empirical hypotheses, but to the artificial language that people might speak if in daily life they embraced the assumptions underlying the puzzle. But the paradox, which was devised by Nelson Goodman,[8] and the fact that it has inspired so much close examination, help to demonstrate the importance of having a theory of conditionality that better expresses the logic of conditionals of ordinary language than the one afforded by standard formal logic.

The puzzle comes to this. Many emeralds have been examined and they have all turned out to be green. We consider ourselves then to have strong evidential support for the truth of the generality 'All emeralds are green'. But now take the predicate 'grue' such that a thing is grue if it is either examined before some arbitrary future time *t* and is green or it is not examined before *t* and is blue. Let 'E' stand for 'is an emerald', 'G' for 'is green', 'T' for 'is examined before time *t*,' and 'B' for 'is blue'. Then the hypothesis that all emeralds are green may be expressed by (I) 'All E are G' and the hypothesis that all emeralds are grue' by (II) 'All E are GT or B–T'. Although (I) and (II) are not inconsistent with each other, when joined with the examination of an emerald after time *t* they imply inconsistent consequences and in this respect are incompatible. But it is claimed that what confirms (I) also confirms (II) and that our evidence for the truth of (I) confirms (II) to the same degree—and on a Nicodian analysis this is true. We have accumulated evidence of the form (E . . .)(G . . .) for the truth of (I), all of which has been gathered before *t*. Therefore, all our evidence also has the form (E . . .)(G . . . T . . .) and in Nicodian terms supports 'All E are GT'. But the evidence form (E . . .)(G . . . T . . .) truth functionally implies (E . . .)(G . . . T . . . or B . . . –T . . .) and so the Nicodian evidence we have for 'All E are G' supports 'All E are GT or B–T' with the same strength.

Is this really a paradox? The factual situations from which we select events in support of the hypotheses in which we have confidence will also contain elements that support countless other hypotheses which, when conjoined with future events, may imply consequences incompatible with those implied by the hypotheses we happen to adopt. On one selection from existing evidence we may be encouraged to assert that all As are Bs while another selection, which we have ignored, would suggest that although all As that are not Cs are indeed Bs, all As which are Cs are not Bs. And future events may show us which hypothesis is the more reliable. So why does (II) alarm us? Is it not merely a dramatization of a general problem? A. J. Ayer has said,[9] 'The underlying point, whatever way we put it, is the simple one made by Hume, that however many As have been found without exception to be B, it remains an open question whether this will also be true of any further As'. Yet we have the feeling that, although Hume's point that hypotheses for which we have direct evidence may ultimately be disconfirmed is well taken, this is not the problem with (II). The problem with (II) is that we rebel against having to accept it on the ground of evidence upon which we base our acceptance of (I). Is there a rational justification for this feeling?

I suggest that there is and that it consists in the fact that if we are bound to accept the 'grue' argument, much of the point of relating our empirical claims to a study of what happens in the world is lost. The argument says that the evidence we adduce in support of a universal conditional A also supports with equal strength another universal conditional which, if true and if events march forward in their accustomed way, will falsify A. Hume says that the evidence for A does not prove it; the 'grue' argument says that the evidence for A is equally evidence for another hypothesis the truth of which would issue in the falsification of A. Let us then return to the beginning and try to see if we are in truth committed to the consequences the argument seems to thrust upon us. The existing Nicodian evidence confirms the claim that all emeralds are green and that they have all been examined before *t*. Because we have evidence of the existence of minerals that are not green and not emeralds, however, but no evidence that there are minerals not examined before *t*, which are not emeralds, we have substantive OC evidence in favor of the assertion that all E are G but no substantive evidence for the claim that all E are T. Observe also that although the Nicodian evidence for 'All E are G' confirms 'All E are G or B–T', the OC confirmation forms of the former do not truth functionally imply those of the latter, which are to be derived from the disconfirmations of 'All E are G or B–T', that is, E . . . –G . . . –B . . . and E . . . –G . . . T . . . . Thus in OC, the confirmations we have for 'All E are G' do not confirm 'All E are GT or B–T,' nor does the former entail the latter.

We should notice that the predicate 'before *t*', in the context of the puzzle, fails to meet the application requirement in that although we are allowed to gain evidence of things examined before *t*, one (logically) cannot, within the constraints laid down by Goodman, produce for inspection anything *not* examined before *t*. It is the odd use that the puzzle makes of this predicate, it seems to me, that distracts our attention from the true villain of the piece: the standard view that 'If A then B' entails 'If A then B or C'.

Chapter 6

THE LOGIC OF LOGICAL STATEMENTS

I

It is the object of this chapter to examine the logical relations of the statements of logical theory. This examination will lead us to study the significance of certain standard logical notions and it will draw attention to the contrasting functions of, on the one hand, the twin concepts of formal validity and contradiction and, on the other, those of logical necessity and impossibility. Ultimately I will present a formal treatment of the logical relations among those statements that belong to the class of what I will call 'logical statements' and that can be described in the following way. There are certain expressions, to be called logical 'generalities' or 'formalities,' which describe the logical characteristics of the forms of primary language. Letting 'A', 'B', . . . stand as blanks for such forms of primary logic as 'p v q', '–r', '–p v p', and also the sentences that interpret these forms, logical formalities will be those expressions that attribute characteristics to those forms, for example, 'A entails B', 'A is consistent', 'B is formally valid'. Some formalities may be seen to be true or false—'A.–A is inconsistent' is true, its negation false—but I shall treat them in general as 'open' sentences that may become true or false when 'A,' 'B,' . . . are replaced by particular primary sentences or by forms of sentences. Thus, although 'B is consistent' is not to be adopted as a statement that is either true or false, we can see that it may be made into a true statement (by replacing 'B' by 'Snow is white' or by 'p v q') or a false statement (if 'B' is replaced by 'Snow is white and not-white' or 'p.–p'). This thought is expressed by saying that 'B is consistent' is possibly true—it can be shown to have some true instances. 'A v –A is valid', however, can be seen to be necessarily true in that all instances in which 'A v –A' is replaced by primary forms produce a valid result. Formalities are thus receptive to modal predicates and will be treated as related to truth and falsity in a modal way. They will be said to be (logically) necessarily true or possibly true or necessarily false. It is these modal statements about logical formalities that I will call 'logical statements'. Instances of formalities will be called their 'exemplifications': 'p v q is consistent' is an exemplification of 'A is consistent'. Formalities that may also be seen to be true, for instance, 'A entails A', will serve as exemplifications of themselves. If we employ the capital letters between 'R' and 'V', both inclusive, to symbolize the formalities, then the logic of logical statements can be said to be about the formal relations among such phrases as 'S is a logical truth', 'If R then necessarily S', 'V is logically impossible—these in turn to be interpreted by putting logical formalities for 'S', 'R', and so on: " 'A' is valid' is a logical truth," "If 'A entails B' then logically necessarily '–B entails –A,' " " 'A.–A is contingent' is logically impossible". It is toward a theory of the relations among these logical statements that the present chapter is largely devoted. But on the way to this destination I will defend the assumption, upon which the account of entailment in previous chapters depends, that although it is on the basis of the contradictoriness or validity of schemata that we judge the entailment relations between their components, the contradictions and valid schemata are not statements and do not function on the left or right sides of entailments.

This chapter presents a sketch only of a theory of the logical relations of logical statements; and although OC diverges from more familiar systems in its approach to entailment and conditionality, I will not attempt to furnish a comprehensive examination of the respects in which the system presented here differs from standard ones. It will be suggested, however, that in many ways the present system is not incompatible with, but offers a redescription of, the deductive procedures of other systems. (See, e.g., the reference to the deductive practices followed in *Principia Mathematica* at p. 170.) I also indicate (section 6) that it is not always possible directly to compare the meaning '. . . is logically necessary' and '. . . is logically possible' have in this essay with the meaning of the standard modal operators '□' and '◇' because they often apply to expressions of different logical types.

My object is to present in outline a logic of the language we use in the construction of first order logics. Although the first order logic developed in the earlier chapters of the essay disallows the entailment, for example, of statements by contradictions, it is not my purpose here to say that this restriction must be a feature of every proper first order logic. My aim is to draw a picture of the logic of the language in which theories of first—order logic are couched, not to define the necessary substance of any such logic. In my view even first order logics that accept that tautologies and contradictions may entail and be entailed do not state the rules of such entailments in laws that are themselves tautologies. And the logic of logical statements that evolves in this chapter does not express its findings in tautologies. But this logic provides a structure within which different opinions of the role of tautologies and contradictions in an object language may be voiced.

Let us try to locate, first of all, the areas of application of the concepts of logical necessity and possibility within the structure of natural language. Statements differ in what they say, are true or false under different circumstances, if the structure of their support and infirmation is different. Two statements say the same thing if their support and infirmation have the same form and are of equal strength. This is why knowledge of the logical relations between statements rests on knowledge of the forms of the statements which would express the evidence for or against them. Scrutiny of the concepts of support and infirmation also permits us to distinguish between different types or levels of statement. Factual statements are strengthened, or undermined, by statements about the nonlinguistic world, statements whose acceptability or otherwise is referable to occurrences of events. Let us call factual statements for and against which evidence is gathered in this way Level I statements.

I shall designate as Level II statements those statements about the environment of assumptions and rules, some varieties of which we observed in chapter 4, which may be said to underlie the language of fact. These commonly accepted rules, and assumptions about the contexts within which factual statements are made, make it possible for us to say what are the pertinent forms of support and infirmation of Level I or factual statements. Level II statements are addressed to commitments and assumptions concerning the application of linguistic expressions imposed by common acceptances of linguistic practices and of the boundaries within which confirmations and disconfirmations are to be sought.

Level II statements cannot be related to Level I statements by way of entailment; the support and infirmation of the two levels belong to different orders. Thus where A is a factual assertion, from 'A has conventional meaning' we cannot deduce the truth of A as a factual possibility. The factual statement A and its accepted meaning are related in such a way that certain linguistic things must be true about the expression that is to carry the assertion if A, used as a factual statement, is to be true or false. The things that support the truth or falsity of a factual assertion and the things that support the truth or falsity of statements about the characteristics of the expressions contained in the factual assertion belong to different conceptual categories. The truth of A is supported or not by the events to which A has reference; the truth of 'A has conventional meaning' is supported by the express or implied agreements with regard to the application of the expressions contained in A by users of the language. Because a factual statement is true only within a system of meanings, one can say that if A is true, then 'A has conventional meaning' is true. But this is not an entailment, because if the truth of A had this relation to the truth of 'A has conventional meaning,' then so would the falsity of A have that relation and we would be bound to conclude that 'A or not–A' entails 'A has conventional meaning'. If this entailment were true, however, then its contrapositive would also be true and we would have to allow that 'A lacks conventional meaning' necessitates the truth of 'A and not–A'. This latter cannot be a logical truth. It is better to say that Level I statements presuppose Level II statements in that the truth or falsity of the former is made possible by the rules and assumptions that are the subject matter of the latter. I believe it is appropriate to treat contextual assumptions as being also related to factual assertions as presuppositions. They are not entailed by the assertions themselves but rather indicate the limits within which support or infirmation of the assertion is to be sought. Level I statements are about what happens, whereas Level II statements are about the prescribed or permitted uses of, or assumptions about, the expressions employed in Level I statements. Evidence for a statement at one level cannot be evidence for a statement at the other level. Evidence for Level II statements justifies particular ways of applying Level I statements; it does not tell us that these applications are supported by circumstances in the world.

I have argued in earlier chapters that, to be true or false, a factual statement must be formally contingent—this is part of the application requirement. If an assertion is to satisfy the requirement, it must be possible to indicate what evidence would support it and what undermine it, and this cannot be done if the assertion's formal structure is that of a tautology or a contradiction. If it is a contradiction it will have no consistent confirmation and if a tautology no consistent disconfirmation.

Formal contingency is equally important to the truth or falsity of assertions about the rules and assumptions that underlie factual statements. These Level II statements are fulfilled or rejected by common acceptances or rejections of usage; but there is no act that is the acceptance of a contradiction, nor any that is the rejection of a tautology, because one cannot say what these acts would be acceptances or rejections *of*. Meaning assertions that have contradictory fulfillments or rejections are not statements. Thus, while arrangements of rules and assumptions help make factual language possible, formal arrangements are necessary to the existence of the languages of both fact and meaning. And as I have called the statements by which we keep track of meaning arrangements Level II statements, the statements by which we indicate the formal structure that supplies the framework for Levels I and II statements will be called Level III statements. These are the expressions (to be called 'logical statements') which purport to say true things about the formal relations between statements of a language. It is to the logical relations of these statements that the present chapter is devoted.

We argue for the truth or falsity of Level I statements on the basis of what we take to be their empirical support. Level II statements show us what is to qualify as support for Level I statements and the truth or falsity of Level II statements is to be estimated in terms of the support and infirmation to be found in the common acceptances and commitments to uses of speakers. Level III statements are concerned with the formal structure of statements of other levels and in particular with the limitations imposed on speech by the need to avoid inconsistency. Inconsistent communications say nothing and it is part of the business of logic to show us how to identify these self-defeating forms and remove them from our speech. Logical statements must then be of crucial importance; without them we will not be able to say which forms are capable of being truc or false. But how do we decide whether logical statements themselves are true or false? It is clear that we cannot appeal to the facts or to accepted usage: responses to the effect that formulations are inconsistent if people in fact treat them as such or if they have agreed to treat them as inconsistent both beg the question, which is "How does one *justify* the description of a formulation as inconsistent?" This is a problem to which I return later in the chapter; but let me here offer one further negative answer. The support of the claim that a logical statement is true—for instance, that 'A is inconsistent' is a truth of logic—is not that its denial is a formal inconsistency. No doubt inconsistencies may be formed from logical sentences. But although, for example, 'This type A statement is consistent and it is inconsistent' is an inconsistency, logical theories are not constructed from denials of assertions of this sort. 'This type A statement is inconsistent or it is consistent' is not a substantive assertion of logical theory. Logical theories say nontautologous things about the concepts of tautology, contingency, validity, contradiction; and the denials of what the theories say, as we shall see, are not formal contradictions.

II

How do theories of logic go about their task of showing us how to escape contradiction? And what role does entailment or deduction play in this undertaking? If our theories about the world are to be of use to us, we must know what conclusions may be drawn from them and we must also know when, on the basis of agreed facts, we can say that a theory has been overturned or made less likely. If we do not know what a theory says about its subject matter, or the circumstances under which it stands corrected, our theories can be neither applied nor rectified. The study of entailment is the study of the limits within which the entitlements to employ or amend a theory take effect.

Let us imagine a theory H consisting of a conjunction of singular and general assertions about the world and suppose that C is a singular or general assertion. When does the entitlement to apply C, on the basis of H, or amend H on the basis of –C, arise? One says that if H entails C, then the assertion of H and –C is inconsistent and further that if H is true and H entails C, then C is true and if H entails C and C is false, then H is false. Thus if one cannot consistently maintain both H and –C, then if H is true one preserves consistency by concluding that C is also and when C is false, consistency can only be preserved by the denial of H.

It follows from this that if (i) H entails C and (ii) C entails D, then (iii) H entails D (H.–D is inconsistent). Because if H is true, the situations in which D is also false—H.C.–D and H.–C.–D—are inconsistent. Similarly, if –D is true, inconsistency will not be avoided unless H is false, because again the alternatives, H.C.–D and H.–C.–D, are contradictions. And to say of some Level III formality, for instance, 'H entails C', that it is a *necessary* truth is to say not only that all instances of 'H entails C' are true but that the denial of one of these instances, although, as I have noted, not self-contradictory, is unintelligible, incapable of rational interpretation. If it is part of the object of theories to say something true, and if inconsistency prevents us from saying anything either true or false, then entailment, the device by which inconsistency is avoided, is important. It is therefore also important that one's theory of entailment be right—hence the value of theories of logic.

There are several things to notice about this account. First, if H is to entail C, it is not essential that everything in H be relevant to the truth of C. Theories contain many assertions not all of which are relevant to their individual logical consequences. If we insisted that each component of an antecedent must participate in all its entailments, we would not be able to deduce (A) from (A.B) or (B) from (A and C and if A then B). Secondly, H must not be inconsistent and C must not be the negation of an inconsistency. If H is inconsistent, then so is the conjunction of H and every arbitrarily chosen –C so that the deductive enterprise collapses; and if the inconsistency of H.–C derives from the inconsistency of –C, then also the point of joining H and –C is lost because it becomes impossible to ascertain the logical effect of –C on H. Thus the components of H must not contradict each other, otherwise H will not be subject to interpretation by a complex statement that risks falsehood, which hazards incompatibility with other statements. Similarly C must not be logically trivial. H and –C must each be purged of inconsistency before the deductive operation can be performed.

In fact one must go further and say that the success of the deductive enterprise presupposes that both H and C are contingent. Because if H is a tautology, then H.–C will only be inconsistent if –C is inconsistent and therefore if C is a tautology. And if C is inconsistent, then H.–C will not be contradictory unless H is inconsistent. Thus neither the assertion nor the denial of H or C may be inconsistent if the assertion of H is to have any useful consequences for our assessment of C and the denial of C for the assessment of –H.

Thirdly, the theory founders if one accepts the equation of the factual conditional with the denial of its disconfirmation. In such case, as we observed, 'If A then B' is deducible from '–A' because the assertion of the latter and denial of the former results in the inconsistency '–A.A.–B'. It has been part of the object of the discussion of conditionals in this essay to show that, when the relations of the statements that confirm and disconfirm them are properly understood, the assertion of '–A' and the denial of 'If A then B' form a consistent conjunction.

One objection that may be directed against this sketch of entailment is that true entailment does not permit the participation of irrelevant elements: that in an entailment, each component of the antecedent must contribute to the deduction of the consequent. Another objection will be that the insistence on the contingency of both H and C makes it impossible to carry through certain essential logical operations, such as proof by *reductio ad absurdum*.

E. J. Nelson[1] held the opinion that the effect of 'p and q' was not simply to express the aggregate truth that 'p' is true and 'q' is true but that it is a unit or whole that expresses the joint force of 'p and q' and that 'p and q' does not entail 'r' unless 'both p and q *function together* in entailing r'. The reason why I do not think that this is correct has been indicated already. The point of having a theory of entailment is to enable us to say to what particular assertions about reality we are committed by all the theories we hold. It defeats the purpose of this enterprise to say that given a conjunction H and C, one cannot say whether or not he is bound to accept C on the ground that not all the components of the conjunction have been used in the derivation. For instance, if A.B really does not entail A, then if, on examination of the facts to which A refers, it turns out that A is false, one cannot, on Nelson's account, say that A.B has been confuted. Though A be false, one is bound to allow, if A is not entailed by A.B, the possibility of the truth of A.B.

What Nelson says about entailment, while not, I think, true of that relation, is true of logical conditionality. The conjunctive components of the condition clause of a true conditional do 'function together' in entitling us to assert the conclusion. Thus, although if 'A' entails 'C', it is true that the conjunction of 'A' and 'B' entails 'C', 'If R and S then necessarily T' cannot be deduced from 'If R then necessarily T'. From the fact that one formality is sufficient to justify the assertion of a second formality, we cannot conclude that any randomly chosen third formality will 'function together' with the first to justify the second. It is a feature of the conditional, at each level of discourse, that all the components of its condition must be relevant to the assertion of the result. Part of what is said by 'A entails B' is that it would be contradictory to assert A and deny B. But 'If R then necessarily S' makes the different claim that, although 'R.–S' is formally consistent, because of the rational connection between 'R' and 'S' if 'R' is the case, one is compelled to conclude that 'S' is the case.

Thus we say that if 'A entails B' is true then logically necessarily '–B entails –A'. But we do not deduce from this that if 'A entails B' is true and if 'A.B does not entail C' is true, then logically necessarily '–B entails –A' is true. The former is a correct statement of a logical law, the latter not. If the latter were true then it would also be true that if 'A entails B' and '–B does not entail –A' were both true, we would have logical justification (which we would not) for saying that 'A.B entails C.'

The second anticipated objection to the account is that the requirement that the antecedent and the consequent of a deduction both be contingent imposes too strict a restraint on the process of deduction. Referring to *reductio ad absurdum* arguments, Jonathan Bennett has said, 'The crucial point is that we do sometimes work with a set of premises S whose modal value is unknown, and learn that S is impossible precisely by finding that it entails (p&–p) or something else which is logically impossible.'[2] Thus, it is reasoned, if we insist that an inconsistent antecedent S is entailed by and entails nothing, then it does not entail (p&–p) and we are preempted from using *reductio ad absurdum* as a method for showing that S is inconsistent. When one reflects on this argument, however, it becomes puzzling. If a consequent C is the inconsistent (p&–p), then it is only entailed by hypothesis H if H.–C, that is, H(p v –p), is inconsistent, which means, because (p v –p) is a tautology, only if H is inconsistent. Thus it seems that one cannot show that H entails C, in these circumstances, unless H is shown to be inconsistent; yet one understood that it was the function of *reductio ad absurdum* arguments to demonstrate that it is the entailment of contradictory C by H that shows us the inconsistency of H.

But *reductio ad absurdum* arguments are not actually used in this way. We apply a *reductio* argument, I suggest, when we conclude that, because A entails B and C entails –B, A, and C are inconsistent. Suppose, for example, we take as given the three assumptions,

(1) (x)(–Fx v Gx),
(2) (x)(Fx v Hx), and
(3) (x)(–Hx v Fx),

and we wish to show, in conventional predicate logic, that (∃x)

(Fx.Gx) is deducible from them.[3] From (1), (2), and (3) there follow, by universal specification, the three forms,

(4) –Fa v Ga,
(5) Fa v Ha, and
(6) –Ha v Fa.

From these, by truth functions, we derive

(7) Fa.Ga.

Because (7) follows from (1), (2), and (3), one way of showing that (∃x)(Fx.Gx) follows from the assumptions is to demonstrate that this consequence of the assumptions is inconsistent with a consequence of the denial of (∃x)(Fx.Gx), (x)(–(Fx.Gx)). If that should prove to be the case, to preserve consistency, one must conclude that (∃ x) (Fx.Gx) is true. It happens that from

(8) (x)(–(Fx.Gx)),
(9) –(Fa.Ga)

is deducible and this is the contradictory of (7), which has been deduced from (1), (2) and (3). What is being said here is that assumption (8) (call it A) entails (9) (say B), and that (1), (2), and (3) (which we call C) entail the denial of (9), which is –B. When a consistent form A entails another consistent form B and a consistent form C entails the denial of B, we conclude that the conjunction of A and C is inconsistent. I believe that this is the manner in which the *reductio* argument usefully functions. It is an argument that derives entailments from individually consistent antecedents which, because they entail inconsistent conclusions, are said to be inconsistent with each other. At no step in the argument is the form A.–A or its denial an antecedent or consequent of an entailment.

This form of the *reductio* argument is the direct result of the principles of transitivity and contraposition for entailment. By contraposition, B entails –C and, because A entails B, and B entails –C, A entails –C. Thus, on the view that transitivity and contraposition are essential characteristics of entailment, any theory of entailment for which exceptions to the reduction argument could be produced would be wrong.

III

I argued in chapter 4 that it is not sufficient, in justifying the claim that an expression has a prescribed use in a language, to say that this is simply the way one does use it. We justify the use by saying that it is one that is accepted as the prescribed use by the speakers of the language, a use to which they are committed. But not even commitment to a usage will justify the claim that a particular procedure is logically necessary. The statement, for example, that 'p.–p' is consistent would not become a logical truth by virtue solely of our common acceptance of it. One cannot answer the contention of A. R. Anderson and N. D. Belnap[4] that the deduction of 'q' from '(p)(–p v q)' is invalid by saying, 'But we all agree that it is valid'. Nor does the blind or unreflective following of a procedure justify the claim that this procedure expresses a logical necessity. And it is possible that Wittgenstein was of this view and that he concluded that because blind following does not justify the assertion of necessity, such assertion is not susceptible of justification. He says, for example, with reference to the question as to how one knows that a mathematical calculation is right, 'The danger here, I believe, is one of giving a justification of our procedure where there is no such thing as a justification and we ought simply to have said: *that's how we do it*'.[5] But although we cannot prove that we must do things in a certain way, we can and do advance (inconclusive) arguments in support of such claims.

This point may be assisted by an example. In section 2 I supported an approach to entailment and in doing so talked about the application of concepts such as entailment and inconsistency in ordinary speech. The statements I have made—for example, that it is insufficient to define entailment of B by A exclusively in terms or the inconsistency of A.–B—do not purport to be formal entailments and so their truth of falsity cannot be established by the tests for entailment that settle questions of the formal relations between Level I statements. Nor, as I have indicated, are they factual or normative claims. The issue is not whether we do observe, or have agreed to accept, particular formal connections. Then what *is* said, for instance, by the claim that A entails B if and only if A.–B is inconsistent? This says, I believe, that the equivalence is a logical necessity, that there are no intelligible exceptions to it. Suppose one responds to this assertion that if the definition be accepted, then A.–A will entail B because A.–A.–B is a contradiction and so the definition compels acceptance of the principle that a self-contradictory assertion entails any assertion. Well, one asks; What is wrong with that? One answer might be that the point of having a theory of entailment is to show us how to avoid inconsistency and because A.–A is already inconsistent, the use of neither B nor –B in conjunction with it can be said to provide an escape from inconsistency. In saying this one is proposing not simply that an exception can be taken to the principle, but that the exception is intelligible, expressive of a possible position. If the exceptional illustration is acknowledged, we also accept the implication that the definition as first stated cannot be relied upon to apply in every case. But the exception need not be acknowledged; in logic, as in other disciplines, the gates to alternative paths are not shut, or if shut not locked. It might be possible to respond that the force of the counterinstance is apparent only, that it cannot be received as an exception because any view of entailment that permits the exception leads to the denial of other assumptions that are even more central. Thus logical arguments are inconclusive: because to the extent that exceptional instances are not contradictory, it may be that an exception to even the most well entrenched principle will recommend itself in the light of a new insight into or altered perspective on a logical issue. And yet it will also be possible, without contradiction, to reject an exception on the ground that the consequences of its acceptance are prejudicial to other valued principles.

Instances of actual uses of expressions that exemplify or form exceptions to logical principles are the observable aspects of the claims made by the principles—principles that are objective in the sense that they assert that what they say cannot be denied without absurdity. And in the case of exemplifications of logical principles, what we observe, or become aware of, is not that they are simply used or that we have accepted their use. What we see or understand is something about our own relation to these claims—that we can or cannot deny them without making nonsense.

The foregoing passage on entailment is not an argument for or against a particular definition of that concept. It is rather an illustration of the way in which logical discussions proceed. And the reader will have noticed that I have predicated logical necessity and possibility of logical formalities, not the forms of Levels I or II statements to which the formalities apply. I have not, for instance, said of A v –A that it is logically necessary or of A.–A that it is logically impossible. In the present theory, logical necessity and possibility will apply to the formalities of logical theory. And formalities are not such expressions as A v B, A v –A, A.–B. These latter expressions are the things about which formalities have things to say. They are the forms of the first order statements that primary logic is about. Expressions that predicate logical characteristics of these primary forms—for instance, 'A is contingent', 'A v –A is tautologous'—are the logical formalities. And in OC we pronounce upon the logical status of these formalities by saying that they are logically necessary or possible or impossible. Formalities are about identifiable characteristics that primary expressions may be found to have. Statements of logical necessity and possibility are about the intelligibility of formalities and their negations. This usage is narrower than the normal one that I think usually allows us to predicate 'necessary' of first–order tautologies and 'impossible' of contradictions. From the perspective of OC, however, first–order tautologies and contradictions constitute, like other expressions of the logical form of object language statements, the subject matter of logical theory; and it is statements *about* this subject matter that are logically necessary or not. I will continue to apply the words 'tautology' and 'valid form' and 'contradition' and 'inconsistency' to the valid and contradictory expressions, respectively, of the language that is the object of study and will reserve 'logically necessary' and 'logically possible' for application to the logical formalities that attribute logical characteristics to the expressions of Levels I and II. It is these latter assertions, assertions that characterize logical formalities as logically necessary or possible, which form the substance of the logic of logical expressions and to which I refer as 'logical statements'. I do not think it would be misleading to describe the logical statements of Level III as 'metastatements' and to say that we have to do in this chapter with logical relations among metastatements.

We may take notice here of the way in which the primary logic of OC, as indicated at the end of chapter 2 (p. 81), is not complete in the sense of that word that would make entailments in OC correspond precisely with all and only the true entailments that hold in the language of study. OC is a theory about part of natural language and is stated in terms of the logical truth or otherwise of logical formalities. These formalities are not themselves tautologies or contradictions, although they are addressed to these and other forms. Because they are not contradictions, exceptions to them may be stated without inconsistency and it will remain an open question whether we will one day find an exception to a logical principle, now accepted, for which a case can be made and that can be understood in a way that does not undermine other principles that continue to hold our allegiance.

IV

It is not my purpose to make a survey of logics or attempt to assess their relative merits. I propose rather to present in outline a formal structure for logical statements. Although substantive logical arguments are not formally valid, they are subject to formal constraints, just as are factual arguments and arguments about conventional meanings. Thus Level III statements themselves have a formal logic, and it will be the task of section 5 to describe the design of this logic. I will show that although the contents of the support and infirmation of logical statements are unique to Level III statements, they bear structural relations to each other and to the statements they support and infirm that are similar to the relations borne by evidential statements to factual and meaning statements. In this sense the three levels of discourse I have distinguished have a unified logic.

Consider the claims that if (i) A entails B and (ii) B entails C, then (iii) A entails C, and replace (i), (ii), and (iii) by the symbols for formalities R, S, and T, respectively. The form of the argument may be represented as follows:

(X) If R and S, then necessarily T.

Although most of us would regard (X) as true (entailment is transitive in OC), it is clear that the structure of (X) itself is not that of an entailment. R.S.–T it not a formal inconsistency. It is possible to consider the falsity of the claim without contradiction, and indeed T. J. Smiley[6] has canvassed the possibility that entailment is not transitive. But the fact that the claim itself is not valid does not, of course, prevent it from having formal consequences. What are these consequences like? Well, for example, suppose one accepts the truth of the conditional 'If R and S then necessarily T', and also that for some application of R and S, the conjunction of R and S is logically true. Is he not committed to the logical truth of T? I am proposing, in other words, that the following is true:

(Y) The conjunction of 'R and S is logically true' and 'If R and S then necessarily T' entails 'T is logically true'.

When Smiley found applications of 'A entails B' and 'B entails C', which seemed true coupled with applications of 'A entails C', which he thought false, it is because he accepted (Y) that he was obliged to reconsider the transitivity of entailment, to reexamine the acceptability of 'If A entails B and B entails C, then necessarily A entails C'. It is possible, then, that although logicians may be able to justify different views of the relations between logical concepts or formalities, they are bound by the same formal constraints on the shape of their arguments. I believe that this is the case. Disagreements are not disagreements unless we have a common view of the consequences of different arguments. In section 5 I show that (Y) can be treated as an entailment.

Let us observe again that entailment is nonconditional—'If A then B', at least in the employment of it that has been explored in the preceding chapters, does not have the same logical properties as 'A entails B'. In the first place, a formality of the form '. . . entails . . .' may be true even though it is clear that not all the components in the antecedent are relevant to the affirmation of the consequent. For example, 'p.q.r' entails 'r'. The irrelevance of 'p' and 'q' to this result does not concern us. But consider the conditional, 'If p and *q* then r'. If we can show, by the systematic application of the principle of confirmation that the presence of *q* in the conditioning event is irrelevant (if *p* and *r* relate to each other conditionally whether *q* is present or not), then 'If p and q then r' is not true. If circumstances permit thoroughgoing confirmation procedures, irrelevant elements in a conditional claim are methodically excluded. But it is not part of the point of saying that 'p.q.r' entails 'r' that all the components of the antecedent are relevant to 'r'. For this reason, although the conditional 'If p and –r then –q' is deducible from 'If p and q then r', it is not a general truth about entailment that " 'p.–r' entails '–q' " follows from " 'p.q' entails 'r.' " If, for instance, 'p' is substituted for 'r', then while it is true that 'p.q' entails 'p', it is not universally accepted that 'p.–p' entails '–q'.[7]

In the second place, if we form, following the example of conditionals, the 'disconfirmation' of an entailment by conjoining its antecedent with the denial of its consequent, what we hope to find is not the disconfirmation of the entailment, but a contradiction, on the basis of which the truth of the entailment is inferred. A risk factor of a well–formed conditional must not be inconsistent with the remaining element in its disconfirmation; but an entailment is *only* an entailment if its antecedent is inconsistent with the denial of its consequent.

An entailment is a nonconditional assertion about the relations between antecedent and consequent to which the tests of entailment may be applied directly. One need not, as one does for conditionals, seek out confirmation by exposing the entailment to circumstances that satisfy the risk factor. If one accepts a proffered test of entailment, then the claim 'A entails B' is demonstrably true or not, in the sense that one knows in what a conclusive demonstration would consist. It is a nonconditional assertion about the logical form of A and B, and the question of its truth, if answered, is conclusively answered. The effect of the confirmation of an ordinary conditional, however, is in its nature not more than provisional. Thus although it is not bad grammar to say, as we often do, that if 'B is deducible from A' is true, then 'If A then B' is true, it is a usage that can lead to the conflation of two distinct logical relations.

I wish also to stress again the difference between the form A.–A and the form 'A.–A is a contradiction'. The former is a contradiction but the latter is no such thing; it is a logical formality, an assertion, which we can say is logically true, about the logical status of the form A.–A. It is logically true because all its instances—" 'Smith is six feet tall and Smith is not six feet tall' is a contradiction"—are undeniable, inescapably the case. Not all logical formalities are necessarily true: thus 'A.–B is a contradiction' is sometimes or possibly true (when, e.g., A and B are both interpreted by 'p') and 'A.–A is consistent' is logically or necessarily false. And although '–A v A is valid' is a logical truth in bivalent logic, it is not valid in the sense in which –A v A is itself valid; its denial does not result in a contradiction. Contradiction involves the application of a descriptive term and its negation to the same subject matter. No such lapse occurs in the predication of 'is not valid' of –A v A.

In the *Tractatus* Wittgenstein said, 'Propositions show what they say: tautologies and contradictions show that they say nothing' and 'Tautologies and contradictions lack sense'.[8] He also said that a tautology's truth is 'certain'.[9] Yet how can something that 'says nothing' and 'lacks sense' be 'certainly true?' One might restate these remarks compatibly with the present discussion by saying not that contradictions and tautologies lack sense but that they are neither true nor false, whereas statements *about* them may be necessarily true—in bivalent logic it is a necessary truth that '–A v A is valid'; but '–A v A is valid' is not itself valid: its denial is not a contradiction.

Although logical statements, which are about tautologies and contradictions are not as such tautologies or contradictions, they may themselves, in combination, express tautologies or contradictions. In such case they are unusable as statements, although they point to the location of Level III entailments. The principle that if A holds and if A entails B then B holds is a necessary truth. But that this principle is a necessary truth and not a necessary truth is a Level III contradiction and is neither true nor false.

Confusion results if we ignore the distinction between tautologus forms and logical formalities and that between entailments and logical conditionals. If we group together valid forms such as

'–A v A' with logical formalities such as '–A v A is valid' under the heading of 'necessarily true,' and if we also maintain (i) that none of these 'necessary' propositions either entails or is entailed by anything and (ii) that '. . . entails . . .' is another way of saying 'If . . . then necessarily , . . .' then we deprive the most valuable formalities of the capacity to function in logical arguments. Logical claims of the sort "If 'A is valid' is true then logically necessarily '–A is a contradiction' is also true" could not be asserted. But it is with the aid of conditional, subjunctive hypotheses of logical theory that we deduce the consequences of these theories. The following example illustrates the point.

We may say of some primary form, for instance, A.–B, that it is a contradiction. Let us call the formality 'A.–B is a contradiction' S. We may also assert the formality 'A entails B' (call this R). If the word 'entails' in R is treated as meaning 'formally entails,' then this is true: If A entails B (R), then necessarily A.–B is a contradiction (S).' Let us express this conditional by R[S(L)], letting 'L' stand for logical necessity. The denial of this conditional is not a contradiction. One might argue, implausibly but without formal inconsistency, that the entailment of B by A does not involve the contradictoriness of A.–B. But the conditional does express a necesssary truth in that if one has any truth of the form 'A entails B', he is entitled, on logical grounds, to conclude that 'A.–B is a contradiction' is also true. To put this in another way, it is not a contradiction to deny that the formal inconsistency of the conjunction of one assertion and the negation of another is a necessary condition of the entailment of the latter by the former; but the denial does seem to lead to the dissolution of intelligible discourse. It is the description of R[S(L)] as an entailment, and the failure to distinguish between formalities and their subject matters that lead to a misunderstanding of the theory that tautologies and contradictions are neither true nor false.[10]

This is because the formality R itself may on occasion be a logical truth (a circumstance to be expressed by R(L)—when it is 'p entails p v q' or 'pq entails p,' and so on. But suppose that R(L). If one accepts the view that R, being a necessary truth, cannot entail anything and that R[S(L)] is an entailment, then one cannot *use* R (L) and R[S(L)] together in the deduction of S(L). But it is just this sort of deduction that we wish to sustain. We do wish to be able to deduce " 'A.–B is a contradiction' is a necessary truth" from the conjunction of "Necessarily 'A entails B' " and "If 'A entails B' then necessarily 'A.–B is a contradiction.' " (It is shown in 6(9), sec. 5, that R(L) and R[S(L)] in conjunction entail S(L).)

The correct view, it seems to me, is that (1) it is forms that are tautologous or contradictory that lack entailment relations and that (2) logical conditionality is not entailment. Logical formalities, which are addressed to tautologous, contradictory and contingent expressions of Levels I and II, may be related to each other in logical conditionals; and it is the association of possible and necessary formalities, and conditionalized ones, which make up the ingredients of entailments between expressions of Level III. Once these distinctions are made, the objections to the theory that tautologous and contradictory forms are neither true nor false lose their force.

An actual example may make this point plainer. Russell and Whitehead, in *2.04,[11] move from the axiom of association,

(a) p v (q v r) . ⊃ . q v (p v r),

to,

(b) –p v (–q v r) . ⊃ . –q v (–p v r).

It might be objected that this example is a counterinstance to my assertion that tautologies do not entail—that we have here a case in which the tautology (b) is deducible from the tautology (a)—in which (a) entails (b). Does not OC, in saying that tautologies cannot entail, seek to repress this innocent and constructive method of deriving logical entailments? But although OC does not call the movement from (a) to (b) an 'entailment', it agrees with the principal deductive claim that is being made in *2.04. OC accepts that the deductive process exhibited in *2.04 is valid but provides a different description of that process. In terms of OC, what the logicians say, in abbreviated style, is first that (a) is a thesis, and a law of the system, and moreover that it is a necessary truth that (a) is a thesis. So,

(i) it is a logical truth that (a) is a thesis.

Also, (b) is a correct substitution in (a), in that in arriving at (b), a routine has been followed that always achieves a substitution of the appropriate kind, and so,

(ii) it is a logical truth that (b) is correctly derived from (a) by substitution.

And it is a rule of logic that,

(iii) if (a) is a thesis and (b) is derived from (a) by appropriate substitution procedures, then necessarily (b) is a thesis.

None of (i), (ii), or (iii) is a tautology. They are all statements of logical theory. But if we accept them, we are driven, on pain of inconsistency, to accept that

(iv) it is necessarily the case that (b) is a thesis.

What I am proposing is that the premisses and consequent of the entailment expressed by the movement from (a) to (b) are not (a) and (b) themselves, those abstractions of Level I expressions; they are assertions as to the logical truth of certain formalities that have (a) and (b) as their subject matter. If R stands for '(a) is a thesis,' S for '(b) is derived by (a) by appropriate substitutions' and T for '(b) is a thesis,' then the entailment I have in mind has the structure,

R(L)
S(L)
RS[T(L)]
———————
entails
T(L)

What we conclude about

$$-p \vee (-q \vee r) \,.\supset.\, -q \vee (-p \vee r)$$

is that its being a thesis is one of the logical truths. And what entitles us to adopt this conclusion is that it is necessarily true that

$$p \vee (q \vee r) \,.\supset.\, q \vee (p \vee r)$$

is a thesis and that thesishood is preserved by the principle of substitution. Thus, although each of the premises R(L), S(L) and RS[T(L)] is nontautological, their relationship to the consequent is one of entailment: R(L).S(L).RS[T(L)] is inconsistent in OC.

When the process that drives the deduction is perceived in this way, one is able to see that the OC theory that tautologies do not entail is compatible with the axiomatic method which, from a different point of view, might be interpreted as a way of deriving tautologies from tautologies.

We see in section 5 how a logic of the relations between logical formalities can be developed. I willl show how Level III assertions R(L), S(L), and RS[T(L)] participate in deductions and in particular how it comes about that conjunctions of the type R(L).S(L).RS[T(L)] entail T(L).

V

The modal predicates L and K will take as subject matter the logical formalities R, S, T, . . . which are themselves addressed to items belonging to Levels I and II, or symbols for such items. It is the function of logical formalities to attribute to Levels I and II items the characteristics that are relevant to their logic. For our purposes, those objects of which R, S, T, . . . are predicated will take the form of sentences ('Smith is red haired') or symbolic components—a, a[b], (ab v c)—and these will be called 'primary expressions', to be represented by the capital letters A, B, C. . . . A primary expression, say (p v q), is a 'substitution instance' of A in R, and when the primary expression has been substituted for A, the whole becomes an 'exemplification' of R. By substituting (p.–p) for A in the formality 'A is a contradiction', we obtain '(p.–p) is a contradiction'. If a primary expression is an instance of A, it must be an instance of each occurrence of A in R. Thus, if we suppose that R is 'A entails A', any expression that instantiates the antecedent must also instantiate the consequent. When (p v q) instantiates A, the whole '(p v q) entails (p v q' becomes an exemplification of 'A entails A'. Different exemplifications of R may be designated by 1, 2. . . .

An exemplification of a logical formality is an 'observable aspect' of that formality, and it will be assumed that there is a method, which can be applied by anyone, for finding out whether the item in question, for instance, (p v q), (p.q), has the asserted characteristic. This means that formalities predicate of primary items features that those items may be determined to have either immediately or by the performance of a test. One must remember, however, that we may be prevented, by human finitude, from being able to determine whether an item has a logical characteristic. A putative entailment may be so complex and intricate that we will be unable to decide whether it satisfies our entailment test; and a conjunction may be so long that we will not be able to ascertain whether or not it is consistent because we will die before all its elements are accounted for.

I shall treat logical formalities as 'open sentences,' which are transformed into statements when exemplified. If it is possible to exemplify R by a statement that is true, R is logically possible, R(K). If it is not possible to illustrate–R by a true statement, R is logically necessary, R(L). The formality 'A is valid' is possibly true because it has some true exemplifications, for instance, '(–p v p) is valid'. But it has false exemplifications as well, for example, '(p v q) is valid', and is therefore not necessarily true. The general form of support for R(K) is the disjunction

R1 v R2 v R3 v . . .

which represents a finite collection of instances and that is established to be true if one of its disjuncts is true. The truth of R1 v R2 v R3 v . . . proves the truth of R(K). To say of a logical formality that it is logically possible is to say that it can have true exemplifications. And the fact that one cannot produce such an example does not prove that one will never be produced. The form of any finite counterinstantiation of R(K) is

–R1.–R2.–R3 . . .

and this does not prove the denial of R(K), –(R(K)).

R(L) is related to its support in the opposite way, being confirmed by a conjunction of exemplifications that do not prove it and infirmed, and disproved, by any counterinstance, and therefore by a disjunction of such.

It will be said that although it is easy enough to deny universality to certain formalities, such as 'A entails B', 'A v B is valid', and the like, it is impossible to find exceptions to certain others, such as 'A.–A is contradictory', and that it is clear that such forms are self-evidently true and thus irrelevant to say that their necessity cannot be proved. It is in fact just this exceptionless character of certain formalities that we assert when we say, for instance, that 'A.–A is contradictory' is logically true. But it remains true that it is not contradictory to deny a formality and that it cannot be shown that an intelligible counterexample that can be treated as a truth will never be forthcoming. It might be possible, for example, to make out a case for the proposition that 'It is and it isn't' in response to 'Is the movie good?' is an intelligible counterinstance of 'A.–A is contradictory' and therefore that 'A.–A is contradictory' is not an exceptionless truth. It would be difficult, perhaps impossible, to defend such a view. But the view itself is not contradictory. It is certainly true that negative examples have occasionally been advanced against the logical truth of formalities that have seemed very obvious. '–A v A is valid' is a formality which, one might think, can take no exceptional instances. One supposes that if a substitution instance of A is false, the denial of it must be true. And yet exceptions have been proposed. It can be argued, to take one example, that 'The present king of France is bald' is neither true nor false.[12] And the possibility of introducing exceptions to '–A entails (If A then B)' has originated the present system and many others. Exceptions to logical universals, to the extent that they are taken seriously, inspire modifications, large or small, in the logical theories of Level III.

Let us observe two features of K and L. (1) The predicates K and L may be interchanged. An expression that predicates K of a formality may be converted into one that contains L, and *vice versa*, by denying the whole expression, denying the subject matter of the expression and changing the modal predicate. Thus, where '≡' is the sign of mutual entailment or logical equivalence,

R(K) ≡ –(–R(L)), and
R(L) ≡ –(–R(K)).

(2) Also, to deny a K or L statement is to deny that the modal predicate applies to the subject matter. Thus,

R(–K) ≡ –(R(K)), and
R(–L) ≡ –(R(L)).

The definition of entailment between expressions that predicate K and L of formalities conforms with the definition used in previous contexts. Assume that X and Y are simple or complex formulae consisting of expressions of the required type, neither the assertions nor denials of which are inconsistent. Then,

> X entails Y if and only if each Co in the one or more termed disjunction of conjunctions formed from X.–Y is inconsistent.

A conjunction Co is inconsistent if it contains a normal inconsistency. And a conjunction is normally inconsistent if the proof of any provable component in the conjunction disproves the rest of the conjunction and provided that by the time each of its Is is confirmed in each alternative way, the corresponding DR will also have been confirmed in each alternative way.

Some illustrative theorems will serve to indicate the logical characteristics of the relations between nonconditional formalities. The first says that if R is logically necessary, it is logically possible.

6(1) R(L) entails R(K)
Co is R(L).R(–K)

The first test on entailment is satisfied because neither R(L) nor R(–K) is provable.

D(I)	I		DR
	R(L)	⟶	R(K)
Confirm (RL) by R1	R1	TF implies	R1
D(II)	I		DR
	R(–K)	⟶	R(–L)
Confirm R(–K) by –R2	–R2	TF implies	–R2

The converse relation is not an entailment because the proof of R(K) does not prove R(L) and the disproof of the latter does not disprove R(K). The second theorem says that 'R is not possible' entails 'R.–S is not possible'.

6(2) R(–K) entails R–S(–K)

Co is R(–K).R–S(K)

R(–K) is not provable and is disproved by the proof of R–S(K).

D(I)	I		DR
	R(–K)	⟶	–R v S(L)
Confirm R(–K) by –R1	–R1	TF implies	–R1 v S1
D(II)	I		DR
	R–S(K)	⟶	R(K)
Confirm R–S(K) by R2–S2	R2–S2	TF implies	R2

6(2) does not oblige us to say, where R(–K), that R entails S, that is, that an impossible proposition entails any proposition. What R–S(–K) says is that it is not possible for both R and –S to be true of an exemplifying instance, not that the conjunction of R and –S is a formal inconsistency. But it is formal inconsistency which would have to obtain if R were to entail S.

6(3)–(7) are also theorems that characterize relations between nonconditional logical formalities.

6(3) R(L).S(L) is equivalent to RS(L)

6(4) R(K) v S(K) is equivalent to R v S(K)

6(5) R(L) v S(L) entails R v S(L)

6(6)	RS(K) entails R(K).S(K)
6(7)	S(L) entails R–S(–K)

We have found that many statements at Level III that assert that one thing 'follows from' another (or that given one thing then necessarily the other) cannot be treated as entailments. For example,

(i) If A–B is inconsistent, then necessarily A entails B,
(ii) If A entails B and B entails C, then necessarily A entails C, and
(iii) If A entails B, then necesssarily –B entails –A,

all, if true, are logical truths. But they are not, in the sense of OC, entailments because their denials are not inconsistent. We decide that they are true if we are convinced that we cannot manage without them and we become convinced of this when we cannot produce plausible counterinstances. But they are deniable without contradiction. I know of no case in which (iii) has been rejected; but T. J. Smiley[13] has addressed himself to the denial of (ii). He is concerned to show that '(A–A) entails B' is false, and so must resist acceptance of its contrapositive, '–B entails (–A v A)'. He does, however, accept '–B entails (–A v A)(–B)' and also '(–A v A)(–B) entails (–A v A)' and so must reject the transitivity of entailment in order to ward off the undesired result. (From the point of view of OC, one rejects the entailment of (–A v A) by (–B)(–A v A). The latter is contingent, true if –B and false if B, but the former is a tautology and not a proper consequent.)

Anderson and Belnap have denied (i). They perceive[14] it to be a truth of logic that (A)(–A v B) does not entail (B) and so, because the conjunction of the former and denial of the latter is a contradiction, produce a counterinstance to the theory that the inconsistency of (A–B) necessitates the entailment of B by A. (In OC, '(A)(–A v B) entails B' is true, but (i) is only accepted for contingent A and B.)

The point is that logical theories are about the intelligibility of alternative formal structures and are susceptible of reasoned criticism and support; they are not themselves the utterances of formal tautologies or contradictions.

Although the logical form of the statements in (i) to (iii) cannot be expressed by 'R entails S', it is not captured by R–S(K) either. Because 'If R then necessarily S' is not true on the ground alone that R is logically impossible or that S is logically necessary. But R–S(–K) *is* entailed by R(–K) and by S(L), as is shown by 6(2) and 6(7). '(A–A) is consistent' is impossible and this entails that '(A–A) is consistent and (A entails B) is false' is impossible, whereas 'A entails B' does not follow necessarily from '(A–A) is consistent'. And although '(A v –A) is valid' may be a necessary truth, it is not for that reason a logical conclusion that may be drawn from any formality at all, such as '(B) is consistent'.

What I suggest is that the forms (i) to (iii) are what they appear to be, and what I have treated them as being—logical conditionals, having the formal characteristics of the factual and conventional conditionals we have studied in previous chapters. They are confirmed by successful responses to the risk of disconfirmation and disconfirmed by compelling instances of the truth of the condition and falsity of the conclusion. Thus 'If (A–B) is inconsistent, then necessarily (A entails B)' is disconfirmed by a case of the inconsistency of (A–B) and the nonentailment of (B) by (A). It is confirmed by true instances of '(A–B) is inconsistent' and 'A entails B' and true instances of 'A does not entail B' and '(A–B) is not inconsistent'.

The relations between logical predicates and their exemplifications, however, provide greater opportunities for leeway or slippage than is the case for factual predicates. It is possible that a disconfirming case can be understood, seen from a point of view, as not disconfirming at all. For instance, it is possible to produce what many will see to be a disconfirmation of (i) by taking 'p–p' as a substitution instance of A and 'q' as a substitution instance of B. Then 'p–p–q' is inconsistent but 'p–p' does not seem to entail 'q'. It is this type of case that has led some writers, including myself, to modify the definition of entailment embraced by (i). In OC, A and B must themselves be contingent if the inconsistency of (A–B) is to result in the entailment of B by A. In Anderson and Belnap's system, the inconsistency of (A–B) does not yield an entailment unless A 'tautologically entails' B. Yet the response of most logicians to the proffered disconfirmation has been to say that the example is not truly disconfirming; if this rejection of the disconfirming instance results in the divergence of the technical meaning of entailment from the meaning it has in ordinary discourse, they say, so be it. In the short run, it is not an easy matter to say whether or not any such modification in the meaning of a logical predicate is a good thing. As a result, disputes about the truth or falsity of claims about logical necessity are sometimes interminably long, and frequently, in any case, much longer than the disputes about the truth or falsity of factual universals.

Although 'R–S is impossible' does not entail 'If R then necessarily S', the converse of this relation is a theorem.

6(8)	R[S(L)] entails R–S(–K)
	Co is R[S(L)].R–S(K)

The proof of the only provable component in Co, R–S(K), disproves R[S(L)].

D(I)	I		DR
	R[S(L)]	⟶	–R v S(L)
Confirm R[S(L)]			
by R1S1	R1S1	TF implies	–R1 v S1
D(II)	I		DR
	R–S(K)	⟶	R[–S(K)]
Confirm RS(K)			
by R2–S2	R2–S2	TF implies	R2–S2

In D(I), it takes only one alternative confirmation of R[S(L)] to confirm –R v S(L) in its single way; in D(II) the I and DR each has but one form of confirmation.

The reader will have noticed that in the above proof I have expressed the denial of R[S(L)] as R[–S(K)]. I have here simply taken advantage of the convention (illustrated at p. 174 in application to nonconditional forms) according to which a form containing one operator can be transformed into an equivalent form containing the other by denying the whole, denying the subject matter of the conclusion and changing the operator.

R[S(L)] is the equivalent of –(R[–S(K)]) and
R[S(K)] is the equivalent of –(R[–S(L)]).

The other convention (p. 174) concerning the negation of modal forms, and that applies to conditionals and nonconditionals alike, is to the effect that the denial of a modal form is equivalent to its corresponding form containing a denial of the modal operator. Thus,

–(R[S(L)]) is the equivalent of R[S(–L)]), and
–(R[S(K)]) is the equivalent of R[S(–K)]

6(9) is the formal statement of the type of relation discussed at pp. 166–7.

6(9)	R(L).R[S(L)] entails S(L)
	Co is R(L).R[S(L)].–(S(L)).

The proof of –(S(L)), the only provable component of Co, disproves the conjunction of R(L) and R[S(L)].

D(I)	I		DR
	R(L).R[S(L)]	⟶	S(L)
Confirm R(L) by R1	(R1)(–R1 v S1)	TF implies	S1
D(II)	I		DR
	R(L).–(S(L))	⟶	R[–S(K)]
Confirm –(S(L)) by –S2	(R2)(–S2)	TF implies	(R2)(–S2)
D(III)	I		DR
	R[S(L)].–(S(L))	⟶	–(R(L))
Confirm –(S(L)) by –S3	(–R3 v S3)(–S3)	TF implies	(–R3)

The I of D(II) and all of the DRs have only one form of confirmation. In D(I) and D(III) there is only one form of confirmation of R[S(L)] which is consistent with the confirmation of R(L) and –(S(L)), respectively. The reader may satisfy himself that, by a similar proof, it can be shown that R(L).S(L).RS[T(L)] entails T(L).

Other theorems which hold with respect to logical conditionals are,

6(10) R[S(L)].S[T(L)] entails R[T(L)],
6(11) R[S(L)] entails –S[–R(L)] and
6(12) R(K).R[S(L)] entails S(K).

Thus transitivity and contraposition hold for logical conditionals [by 6(10) and 6(11)]; and 6(12) says that from the truth that R is logically possible and that if R then necessarily S, one may deduce that S is logically possible. The traditional paradoxes do not affect us: neither R(–K) nor S(L) entails R[S(L)].

In chapter 3 (pp. 100–101) I drew attention to the fact that in OC 'All f is g and all g which is h is j' does not entail 'All f which is h is j'. The same situation obtains for logical statements.

R[S(L)] and ST[U(L)] does not entail RT[U(L)].

It seems to me that this position too is defensible. It is true that if A is a contradiction (R), then necessarily A.B is a contradiction (S) and also that if A.B is a contradiction (S) and A is valid (T) then B is a contradiction (U). But it is right, I suggest, that OC does not, if we accept these premises, compel us to say that if A is a contradiction

contradiction (U). But it is right, I suggest, that OC does not, if we accept these premises, compel us to say that if A is a contradiction and also valid, then B is a contradiction. Some would say that anything can be deduced from an absurdity; others that nothing can. Whether it is right or wrong, however, the latter view does not represent a formal inconsistency.

VI

This treatment of logical necessity cannot be compared in a direct way with the alethic modal logic founded on the systems of C. I. Lewis. In the first place, the □ and ◇ of standard systems and the L and K of OC are not predicated of the same things; □ and ◇ are typically predicated of the well formed formulas of the sentential calculus, for instance, 'pv–p,' 'p v q.' L and K, however, apply to such expressions as '□p v –p', '◇ p v q'. Thus '□p v –p' is itself a logical formality, upon the logical status of which " '□p v –p'(L)" is a comment. A second difference between the two types of systems is that whereas OC offers a formalization of logical conditionality and distinguishes the conditional from entailment, standard modal logic provides a formalization of entailment and does not, in general, discern substantive differences between entailment and logical conditionality. The differences between 'If A then necessarily B' and " 'A' entails 'B' " are considered to be grammatical rather than logical.

A further difference is that the □ of standard modal logic may be asserted of things the denial of which would be absurd as well as of things the denial of which would be self contradictory. So it is true that '□ If he's a bachelor, he's unmarried' and also that '□–(p.–p),' and as a result it seems that the meaning of □ may vary according to the type of item of which it is predicated. The truth of the first statement depends on the conventional meaning of the predicates 'is a bachelor' and 'unmarried', that of the second on the formal characteristics of any acceptable interpretation of –(p.–p). The L of OC, however, always addresses itself to logical formalities, which are contingent expressions, and it says that the denial of the formalities could not take intelligible exemplifications. The L of OC has nothing to do with the meanings of Level I predicates.

Although OC does not formalize entailment, it does comprehend a distinctive meaning for that concept and it is therefore possible to comment of features of the version of entailment reflected in standard modal logic and the ways in which it differs from the entailment concept of OC.

Like the standard modal rendering of entailment as $\Box(A \longrightarrow B)$ (where $(\ldots) \longrightarrow (\ldots)$ is equivalent to $-(\ldots) \vee (\ldots)$), the OC version is transitive and subject to contraposition (see p. 81). Also, it is not necessary that all the components of the antecedent participate in the entailment of the consequent. It is a requirement of entailment in OC, however, that antecedent and consequent be contingent. But where the OC notion of logical following differs most noticeably from the standard theory is in its introduction of a distinction between entailment and logical conditionality. It is in relation to this latter kind of logical consequence that relevance must be maintained and here that we respect Nelson's insistence that the components of what is supposed must act together in justifying a conclusion. This fact about true conditionals, that the components of their premises must act together in justifying their conclusions, emerged from our attempt to discover the significance of another fact about them—that their confirmations give evidence of, but are not equivalent to, the truth of what they say. And in the course of this effort, we came upon the modified conception of inconsistency that enabled us to accommodate the entailment relations of both conditional and nonconditionals.

It is not my view that the OC account of entailment is the only one which it is possible rationally to take; the OC version does, however, facilitate the theory of conditionals presented in OC and it affords an alternative for those who do not wish to be required to deduce any statement whatever as a formal consequence of a contradiction or to treat randomly selected statements as justifications of tautologous consequents.

We have seen that statements can be classified on the basis of the different types of risk to which they may be exposed. And the three types of statement to which particular attention has been paid are the factual, prescriptive and logical, which severally face the risks of incompatible experience, disagreement as to usage and unintelligibility. The special importance of the normative and logical forms of statement lies in their significance for the language of fact. Factual statements cannot be true or false unless the expressions they contain have agreed uses and unless they are formally consistent. Hence factual statements exist within a structure of conventions and logical necessity. And because we can only express the influence of considerations of conventional meaning and logical necessity on the logic of factual statements by *using* meaning statements and logical statements, it is important to know the logic of these latter types of statement. I hope that in this essay I have succeeded in exposing part of the logical structure of these supportive domains of language and of their relations to the logic of the language of fact.

Appendix

I show in this part that when the four traditional subject predicate forms are treated in the manner argued for in sections 7 and 8 of chapter 3, all twenty-four syllogisms commonly accepted as valid can be proved as entailments in OC. In the syllogism, it will be remembered, a conclusion is said to be deducible from two premises, each of the premises and the conclusion being in one of the four forms A, E, I, or O. It can be demonstrated that each of the twenty-four valid syllogisms is provable as an entailment in OC and that each syllogism is based on a normal inconsistency. This being the case, we should be able to discover eight inconsistencies from which all twenty-four syllogisms can be derived. It happens that these eight normal inconsistencies can be found; they are as follows:

I	a[b(A)].b[c(A)].a[–c(S)],
II	a[b(A)].b[c(A)].a[–c(A)],
III	a[b(A)].b[–c(A)].a[c(S)],
IV	a[b(A)].b[–c(A).a[c(A)],
V	a[b(S)].b[c(A)].a[–c(A)],
VI	a[b(S)]b[–c(A)].a[c(A)],
VII	a[b(A)].b[c(A)].c[–a(A)] and
VIII	a[b(S)].b[c(A)].c[–a(A)].

Let us first establish the inconsistency of I.

I	a[b(A)].b[c(A).a[–c(S)]		
D(I)	I		DR
	a[b(A)].b[c(A)]	——>	a[c(A)]
(i) Confirm			
a[b(A)] by a1b1	(a1b1)(–b1 v c1)	TF implies	a1c1

(ii) Confirm b[c(A)] by –b2–c2	(–a2 v b2)(–b2–c2)	TF implies	–a2–c2
D(II)	I		DR
	a[b(A)].a[–c(S)]	⟶	b[–c(S)]
Confirm a[–c(S)] by a3–c3	(–a3 v b 3)(a3–c3)	TF implies	b3–c3
D(III)	I		DR
	b[c(A)].a[–c(S)]	⟶	a[–b(S)]
Confirm a[–c(S)] by a4–c4	(–b4 v c4)(a4–c4)	TF implies	a4–b4

The formulation satisfies the first test on entailment as well, because the proof of the only provable component, a[–c(S)], disproves the rest of the conjunction I.

The syllogism consists of a major premise containing the major and middle terms, a minor premise containing the minor and middle terms and a conclusion, of which the subject is always the minor term and the predicate the major. If the conclusion is deducible from its two premises, the syllogism so formed is said to be valid. Let us designate the minor, middle, and major terms as s, m, and p, respectively, and express the syllogistic form by setting out first the major premise, the minor premise underneath it, and the conclusion underneath the premises and separated from them by a line. Observe that I is the same conjunction as (Q), (p. 107). Therefore, by substituting in I s for a, m for b, and p for c, and by making the conclusion the denial of the third component, we obtain the syllogism *Barbara*,

m[p(A)]
s[m(A)]
―――――――
s[p(A)].

Baroco is obtained when the conclusion is the denial of the first component and the substitutions are s for a, m for c, and p for b.

p[m(A)]
s[–m(S)]
―――――――
s[–p(S)]

Bocardo results when the conclusion is the denial of the second component and the substitutions are m for a, s for b and p for c.

m[–p(S)]
m[s(A)]
―――――――
s[–p(S)]

Conjunctions II to VIII can also be proved inconsistent; and by a similar process of substitution it is possible to show that the remaining twenty-one syllogisms are obtainable from the remaining seven inconsistencies.

II	*Felapton, Camestrop, Barbari*
III	*Celarent, Festino, Disamis*
IV	*Darapti, Cesaro, Celaront*
V	*Darii, Camestres, Ferison*
VI	*Cesare, Ferio, Datisis*
VII	*Bramantip, Fesapo, Camenop*
VIII	*Dimaris, Fresison, Camenes*

The forms of the syllogism have traditionally been divided into four 'figures' such that in the first figure the middle term is subject of the major premise and the predicate of the minor; in the second it is predicate of both major and minor; in the third figure it is subject of both major and minor; and in the fourth figure it is the subject of the minor and predicate of the major. Because each of the premises and the conclusion will take one of the four forms A, E, I, or O, there are 64 possible syllogisms in each figure and, because there are 4 figures, 256 syllogisms in all. It is possible, therefore, to establish that only 24 forms of the syllogism are valid by applying the consistency test directly to each of the 256 conjunctions formed from the premises of each possible syllogism and the denial of its conclusion.

The same result, however, may be achieved through an analysis of the OC inconsistencies from which the syllogistic deductions are derived. The three terms a, b, and c included in the syllogistic form are organized in systems of three pairs, for example, (ab)(bc)(ac), in which one term is excluded from each pair. All reversals of order in the letters of each pair can be reduced, by relettering, to either

(X) (ab)(bc)(ac) or
(Y) (ab)(bc)(ca).

Through a consideration of the general characteristics that

must be possessed by the traditional subject predicate form and by OC inconsistencies, we can discover the number of inconsistencies of the requisite type that can be extracted from the models (X) and (Y). The characteristics in question are these:

(1) If the association of any two components of a conjunction is to entitle us to deduce the denial of the remaining component, the components must be capable of testing each other in such a way that at least one confirmation of one of the components in an association of two components is an element in the disconfirmation, and disproof, of the other component.

(2) The traditional subject predicate form does not include negative subjects and so no component of a conjunction may employ a negative condition: there is no statement of the form A, E, I, or O, which corresponds to –a[b(A)] or –a[b(S)]. [1]

(3) There may not be more than one particular form (an I or an O form) in an inconsistent conjunction. In such a conjunction, one component in every pair of components must be capable of 'forcing' the other. If forcing does not occur, the denial of the remaining component cannot be deduced. Neither one of two particular forms forces the other.

(4) If a conjunction C2 can be reduced to another conjunction C1 by changing the order of the components and relettering the terms, C1 and C2 will be said to express the same conjunction.

(5) The conjunction must satisfy the OC tests on inconsistency.

Observing these constraints, let us see how many inconsistencies we can make from (X) and (Y).

(X) (ab)(bc)(ac).

(I) (ab) is universal. It may not be universal negative because if it were, either b of (bc) or a of (ac) would be required to be negative and then either (bc) or (ac) could not take one of the four forms A, E, I, or O. (bc) must be universal. If it were not it would not be forced by the confirmation of a[b(A)] and its own single confirmation form would include b, which is not an element of the disconfirmation of a[b(A)]. Then (bc) will be of the form A or E. But if the first two components are to be inconsistent with (ac), then if (bc) takes the form A, (ac) must be E or O, and if (bc) takes the E form, then (ac) must be

A or I. Thus when (ab) is universal, it is necessary that the requisite inconsistency be in one of the four arrangements,

(X)(I) (i) a[b(A)].b[c(A)].a[–c(S)],
(ii) a[b(A)].b[c(A)].a[–c(A)],
(iii) a[b(A)].b[–c(A)].a[c(S)] or
(iv) a[b(A)].b[–c(A)].a[c(A)].

(II) If (ab) is particular, its form must be I. If it were O, its single confirmation form a . . . –b . . . would require us to give (bc) a negative subject. The forms (bc) and (ac) must both be universal (because an inconsistency cannot support more than one particular statement) and, to preserve inconsistency, if (bc) is A (ac) must be E and if (bc) is E (ac) must be A. Inconsistencies containing an (ab) in form I must then be of the type,

(X)(II) (i) a[b(S)].b[c(A)].a[–c(A)] or
(ii) a[b(S)].b[–c(A)].a[c(A)].

(Y) (ab) (bc) (ca)

(I) Let (ab) be A. The form, (bc) will be universal. If (bc) were particular, the confirmation of (ab) would not jeopardize it and the proof of neither b[c(S)] nor b[–c(S)] would place a[b(A)] in a state of risk. If (bc) is A, (ca) will not be particular because the proof of neither c[a(S)] nor c[–a(S)] would disprove a[b(A)].b[c(A)]. And if (bc) is A, (ca) must be E. If (bc) is E (ca) must be A or I. For (Y), then, when (ab) is A, there are not more than three conjunctions which can be inconsistencies and they are,

(Y)(I) (i) a[b(A)].b[c(A)].c[–a(A)],
(ii) a[b(A)].b[–c(A)].c[a(A)] and
(iii) a[b(A)].b[–c(A)].c[a(S)].

(II) Let (ab) be E. If (bc) were particular negative, the inconsistency of the conjunction would demand a negative subject of (ca). But (bc) may take the particular affirmative form I, and if it does, (ca) must be c[a(A)]. In this arrangement, then, there is the single inconsistency,

(Y)(II)(i) a[–b(A)].b[c(S)].c[a(A)].

If (ab) is E and (bc) is universal, (bc) must be affirmative. Were it not, a negative subject would be required of (ca). The arrangement produces the inconsistency,

(Y)(II)(ii) a[–b(A)]b[c(A)].c[a(A)]

(The conjunction a[–b(A)].b[c(A)].c[a(S)] is not inconsistent because the single confirmation of c[a(S)] does not contain an element of the disconfirmation of b[c(A)].)

(III) Let (ab) be particular. It must be affirmative if (bc) is not to have a negative subject. The forms, (bc) and (ca) must be universal, and (ca) will have a negative subject if (bc) is universal negative. Therefore (bc) will have the form A and, to obtain the inconsistency, (ca) will be E. Thus for (Y)(III) we have just the inconsistency

(Y)III)(i) a[b(S)].b[c(A)].c[–a(A)].

Now observe that by reordering the components and relettering them, (Y)(I)(ii) and (Y)(II)(ii) reduce to (Y)(I)(i). Thus a[b(A)].b[–c(A)].c[a(A)] becomes a[b(A)].b[c(A)].c[–a(A)] if we make c[a(A)] the first component to yield,

c[a(A)].a[b(A)].b[–c(A)]

and then replace c by a, a by b, and b by c. And a[–b(A)].b[c(A)].c[a(A)] becomes (Y)(I)(i) if we make b[c(A)] the first component and c[a(A)] the second to yield,

b[c(A).c[a(A)].a[–b(A)]

and then replace b by a, c by b, and a by c.

Also, (Y)(II)(i) becomes (Y)(III)(i) if we make b[c(S)] the first component and c[a(A)] the second to produce,

b[c(S)].c[a(A)].a[–b(A)]

and then replace b by a, c by b, and a by c. Finally, (Y)(I)(iii) also becomes (Y)(III)(i) if we make c[a(S)] the first component and a[b(A)] the second to obtain.

c[a(S)].a[b(A)].b[–c(A)]

and then replace c by a, a by b, and b by c.

The method shows us that there are not more than eight inconsistencies of the required type:

(X)(I) (i) a[b(A)].b[c(A)].a[–c(S)],
(ii) a[b(A)].b[c(A)].a[–c(A)],

(iii) a[b(A)].b[–c(A)].a[c(S)],
(iv) a[b(A)].b[–c(A)].a[c(A)],
(X)(II) (i) a[b(S)].b[c(A)].a[–c(A)],
(ii) a[b(S)].b[–c(A)].a[c(A)],
(Y)(I) (i) a[b(A)].b[c(A)].c[–a(A)] and
(Y)(III) (i) a[b(S)].b[c(A)].c[–a(A)].

Thus there are at most eight inconsistencies that support valid syllogisms and these are the same inconsistencies as (I) to (VIII), from which we derived the twenty-four commonly accepted syllogisms of the traditional logic.

Notes

Preface

1. For a review of contemporary theories of conditionals, see Donald Nute's essay, "Conditional Logic," in *Handbook of Philosophical Logic*, eds. D. Gabbay and F. Guenther, vol. II: (Reidel).

Chapter I

1. The example is a modified version of one given by R. Stalnaker in his essays, "A Theory of Conditionals," first printed in *Studies in Logical Theory*, ed. N. Rescher (Oxford: Blackwell, 1968), reprinted in *Causation and Conditionals*, ed. E. Sosa New York; Oxford U.P. 1975.) It is aj type of example that makes it difficult to see the singular conditional's potential for taking alternative forms of confirmation, a potential that I argue is intimately related to its logic.

2. H. L. A. Hart and A. M. Honore, *Causation in the Law*, p. 52. (New York: Oxford University Press, 1959).

3. Stalnaker. "A Theory of Conditionals," *op. cit.* David Lewis, *Counterfactuals* (Oxford: Blackwell, 1973).

4. Stalnaker, "A Theory of Conditionals," p. 169.

5. The Stalnaker and Lewis theories differ in some respects, but they resemble each other on the points discussed.

6. Lewis, *Counterfactuals*, p. 28 (italics mine).

7. Ibid.

8. This is Stalnaker's account. The Lewis version differs in a way that does not matter for the fate of contraposition.

9. Lewis also makes use of an 'even if' conditional in illustrating the 'possible worlds' approach to contraposition. His example is, 'If Boris had gone to the party, Olga would still have gone,' which does not assert that Boris's presence at the party would entitle us to conclude that Olga will be

there. It rather states that if Boris had gone to the party, this would not have justified us in concluding that Olga would not have attended. See Lewis *Counterfactuals*, p. 35.

10. Wayne Davis has introduced a system, based on considerations of possible worlds, in which for a certain type of conditional, called the 'strong conditional,' contraposition is valid ("Weak and Strong Conditionals," *Pacific Philosophical Quarterly* 64 (1983): 57–71). The weak conditional, A ⟶ ⟶ C, which Davis says lacks a 'does not assert a connection between condition clause and conclusion. 'If you flip the light switch, you will not thereby turn on the radio,' only implies that flipping the light switch is not connected to turning on the radio. The strong conditional, A ⟶ ⟶ C, to be expressed with the use of 'then,' implies that a connection does obtain between the events represented by the condition clause and the conclusion clause. In Davis's system the weak conditional A⟶C is true if and only if C is true in i(A) where i(A) is the closest possible world in which A is true. And the strong conditional A⟶ ⟶ is true if and only if A⟶ and C⟶ A are both true, which is to say, if C is true in i(A) and –A is true in i(–C). One might think that where there is no connection whatever between the condition and the projected conclusion the contrapositive of a weak conditional will often be true. When, for example, (1) 'If you flip the light switch, you will not thereby turn on the radio' is true, under the same circumstances (2) 'If you turn on the radio, it will not be because you have flipped the light switch' will also be true. On this analysis, Davis's proposal becomes very doubtful: (1) does not become a strong conditional just because the two weak conditionals (1) and (2) are both true. Davis, however, gives the contrapositive an entirely different reading. He transforms the conditional into its contrapositive in such a way as to make the denial of the conclusion clause involve the assertion of the 'if' clause. Thus he says that 'If you flip the light switch, then you will not thereby turn on the radio,' taken as a strong conditional, is false and its contrapositive is false because (so he says) the denial of the conclusion clause necessitates the truth of the 'if' clause: 'most similar world in which you turn on the radio by flipping the light switch is a world in which you flip the light switch' (p. 63). But this is an extraordinary treatment of contraposition and one that would render all strong conditionals false. For instance, 'If you switch the light switch then you will turn on the light,' understood as a strong conditional, becomes false because the most similar world in which you do not turn on the light by flipping the light switch is a world in which you flip the light switch. The interpretation requires us to say that a world in which the 'then' clause is false is one in which the 'if' clause is fulfilled. In OC, Davis's weak conditional, 'If you flip the light switch, you will not turn on the radio' may be interpreted as the denial of a conditional, as 'It is not the case that if you flip the light switch then you will turn on the radio.' It asserts that one would not be justified in concluding that the radio will come on if one flips the light switch.

11. See, for example, E. W. Adams, *The Logic of Conditionals*, pp. 14–15. (Dordreht/Boston: D. Reidel Publishing Company, 1975).

12. Stalnaker, "A Theory of Conditionals," p. 173.

13. Frank Jackson, *Conditionals* (Oxford: Blackwell, 1987), p. 81.

14. Ibid.

15. Stalnaker, "A Theory of Conditionals," p. 169.

16. Ibid., p. 168.

Chapter 2

1. The reference is to p. 2. On that previous occasion I ignored the distinction made here between 'elements' and 'components'.

2. See my remarks on this feature of 'even if' conditionals in chapter 1, pp. 33–34.

3. J. L. Pollock, *Subjunctive Reasoning*, p. 29. (Dordrecht/Boston: D. Reidel Publishing Company, 1976)

4. See Preface, pp. xiii–xiv.

5. Ibid., p. xi.

6. Thus, the apparently innocent (I) 'pq v p–q[r]' fails because the fulfillment of one of its risk factors results inescapably in its disconfirmation. The form is disconfirmed by the fulfillment of 'p. . . q . . . –r . . .'and also by 'p. . . –q . . . –r . . .' and so fulfillment of the risk factor 'p. . . –r . . .' whether accompanied by fulfillment of 'q. . .' or '–q . . .' must result in the disconfirmation of (I). (I) is, from the point of view of OC, an abortive attempt to say 'If p, then r, whether or not q'. I suspect that this thought—that a certain element is irrelevant to a particular outcome—cannot, at least in OC, be contained within an expression that itself asserts a relation of conditionality.

Substitution of truth functionally equivalent conditions cannot, for these reasons, be relied upon to preserve the truth. The form, 'p[r]' does not entail 'pq v p–q[r]' because the latter, as we have seen, is not well formed. Truth is preserved, however, by the substitution of equivalent antecedents of entailments. For example, if A entails C, then so does AB or A–B. (For an account of the problems created for possible worlds logics by the acceptance as well formed of the schema 'If p and q or p and –q, then r,' see Donald Nute's essay, "Conditional Logic", in *Handbook of Philosophical Logic*, eds. D. Gabbay and F. Guenther, vol. 11, pp. 413–418.

7. An. Pr. 57^{b} 1ff. The Works of Aristotle, ed. Sir David Ross, Vol. 1, Analytica Priora, 57^{b} 1 ff, (New York: Oxford University Press, 1928).

8. See A. N. Prior, *Formal Logic*, p. 170, (New York: Oxford Univer-

sity Press, 1962) and also J. Lukasiewicz, *Aristotle's Syllogistic*, pp. 49–50. (New York: Oxford University Press, 1957). I say more about Aristotle's views *infra*, p. 78.

9. See p. 28.

10. Simon Blackburn, *Spreading the Word* (New York: Oxford University Press, 1984.) The reader may wish to satisfy himself that the same type of suppression of a relevant assumption in part of a premise occurs in the examples by Wayne Davis ("Weak and Strong Conditionals," 64 (1983): 57–71, *Pacific Philosophical Quarterly*), Storrs McCall ("Counterfactuals Based on Real Possible World," Nous 18 (1984): 463–77), and R. C. Stalnaker ("A Theory of Conditionals", in *Causation and Conditionals*, ed. Ernest Sosa, p. 173. (New York: Oxford University Press, 1975). The neglected assumptions are (Davis) that Alan's wife does not attend the party, (McCall) that Casper does not attend the party and (Stalnaker), that J. Edgar Hoover is a U.S. citizen.

11. The Works of Aristotle, ed. Sir David Ross, Vol. 1, Analytica Priora, 57^b 1ff (New York: Oxford University Press, 1928).

12. Lukasiewicz, *Aristotle's Syllogistic*, p. 50.

13. This is presumably an entailment in Stalnaker's theory. If q is true in the closest world in which p is true, q v r is also true in that world.

14. Stalnaker, "Theory of Conditionals", p. 173.

Chapter 3

1. See section VIII (p. 107) for an account of the relations of this theorem to the traditional formal logic. Proof of the inconsistency of g[h(N)].f[g(N)].–(f[h(N)]) is presented in the Appendix.

2. It was known to medieval logicians. See A. N. Prior, *Formal Logic*, p. 117. (New York: Oxford University Press, 1962).

3. See ibid., pt. 2, W. and M. Kneale, *The Development of Logic*, chapter 2. (New York: Oxford University Press, 1962). David Mitchell, *An Introduction to Logic*, chapter 2; (New York/London: Hutchinson University Library, 1964). P. F. Strawson, *Introduction to Logical Theory*, chapter 6. (London: Methuen & Co. Ltd., 1952).

4. E. J. Lemmon, *Beginning Logic*, p. 174. (London: Thomas Nelson and Sons Ltd., 1965).

5. The interpretation is not universally accepted. See "Truth-Functional Conditionals and Modern vs. Traditional Syllogistic", R. B. Angell, *Mind*, Vol. xcv, (1986); p. 210, where a rationalization of the traditional system is set forth in which the A and E forms are treated as conditionals and

O and I as their respective denials. In Angell's system, 'Some S is P' does not imply that there exist things that are S and P; it is simply the denial of 'No S is P,' which is given the conditional form, 'For all x, if x is S, then it is not P'.

6. See Mitchell, *Introduction to Logic*, p. 35.

7. It is true that in OC b(N) entails –a v b(N). Also, from the truth of b(N) and a1, b1 is deducible. The conjunction b(N).a1.–b1 contains the normal inconsistency b(N).–b1. But an entailment is not a conditional.

8. The Works of Aristotle, ed. Sir David Ross, vol. 1, Analytica Priora, 57^b 9ff. (New York: Oxford University Press, 1928).

9. C. Ladd-Franklin, "Symbolic Logic", in Baldwin's Philosophical Dictionary, ed. James M. Baldwin (New York: New MacMillan Company, 1918).

Chapter 4

1. W. V. Quine, *Ontological Relativity and Other Essays*, p. 29. (New York: Columbia University Press, 1969).

2. Quine, "Mind and Verbal Dispositions," in *Mind and Language*, ed. S. Guttenplan, p. 92. (New York: Oxford University Press, 1975).

3. Saul Kripke, *Wittgenstein on Rules and Private Language*, (Mass.: Harvard U.P.), 1982, pp. 26–28.

4. *Ontological Relativity*, p. 29.

5. Quine, "Mind and Verbal Dispositions," p. 89.

6. See pp. 72–74.

7. One of the many things Wittgenstein says about rules is that we follow them 'blindly': *Philosophical Investigations*, section 219 (Oxford: Basil Blackwell, 1955) ("When I obey a rule, I do not choose, I obey the rule *blindly*"). I do not think it possible to state with confidence the respects in which rules in Wittgenstein's account of them resemble conventions in the present theory. For persuasive but conflicting discussions of Wittgenstein's theory of rules, see Kripke, *Rules and Private Language*, Colin McGinn, *Wittgenstein on Meaning*, (Oxford: Basil Blackwell, 1984), G. P. Baker and P. M. S. Hacker, *Scepticism, Rules and Language*, (Oxford: Basil Blackwell, 1984).

8. See the Supreme Court of Canada case of Kilgoran Hotels and Samek. Canada Law Reports, Supreme Court of Canada [1968] p. 3.

9. See pp. 79–80.

10. Frank Jackson, *Conditionals*, p. 85. (Oxford: Basil Blackwell, 1987).

11. See chapter 2, p. 53.

12. Ernest Adams, "Subjunctive and Indicative Conditionals," in *Foundations of Language*: 6 (1970); 89–94.

Chapter 5

1. MIND, 54 Hempel, "Studies in the Logic of Confirmation," 97–121, reprinted in *Aspects of Scientific Explanation* (New York: The Free Press, 1965).

2. This quotation appears in ibid., p. 10, and is taken from the 1930 edition of Nicod's *Foundations of Geometry and Induction*. International Library of Psychology, Philosophy, and Scientific Method (London: Kegan Paul, Trench, Fruhner & Co., 1930). The passage occurs, in slightly different translation, in the 1969 (Routledge and Kegan Paul) edition of Nicod's book at p. 189. (London: Routledge and Kegan Paul, 1969)

3. Nelson Goodman, *Fact, Fiction and Forecast*, chapter 3. (New York: Bobbs-Merrill Company, Inc., 1965 edition)

4. See W. V. Quine, "Natural Kinds," in *Ontological Relativity and Other Essays*, p. 115. (New York: Columbia University Press, 1969)

5. Ibid.

6. See in particular, Theorem 4(6) at p. 135.

7. G. H. von Wright, *The Logical Problem of Induction*, pp. 66, 69. (Oxford: Basil Blackwell, Second Revised Edition, 1965)

8. Goodman, Fact, Fiction and Forecast, chapter 3.

9. *The Central Questions of Philosophy*, (London: Weidenfeld and Nicholson, 1973) p. 178.

Chapter 6

1. E. J. Nelson, "Intentional Relations," XXXIX, *MIND*, (1930): 400.

2. Jonathan Bennett, "Entailment", *Philosophical Review*: 78 (1969): 208.

3. The example is taken from Benson Mates, *Elementary Logic*, p. 119. (New York: Oxford University Press, 1972)

4. A. R. Anderson and N. D. Belnap. *Entailment*. (Princeton: Princeton University Press, 1975)

5. Wittgenstein, *Remarks on the Foundations of Mathematics* (Mass.: M.I.T. Press), p. 98. G. P. Baker and P. M. S. Hacker argue that, for Wittgenstein, to say 'This is simply what I do', with respect to a rule, *is* to justify it, but that no further justification can be given. See their *Scepticism*,

Rules Language. (Oxford: Basil Blackwell, 1984) p. 82. It is not however clear, on the basis of what he actually said, that this is what Wittgenstein meant to say about logical and mathematical rules.

6. T. J. Smiley, "Entailment and Deducibility", *Proceedings of the Aristotelian Society*, n.s., vol. 59, pp. 233–54.

7. That 'p.–p' entails '–q', however, is usually accepted. Anderson and Belnap (*Entailment*, p. 491) reject the entailment; but they treat 'If p then q' and " 'p' entails 'q' " as interchangeable.

8. *Tractatus, Logico-Philosophicus*, 4.461. (New York: The Humanities Press, 1971 edition)

9. Ibid, p. 4. 464.

10. See P. F. Strawson, "Necessary Propositions and Entailment Statements", *MIND*, LVII (1948), 184. For Strawson, (p. 186), '[N]o necessary statement and no negation of a necessary statement can significantly be said to entail or be entailed by any statement' and " 'p' entails q' " means " 'p ⊃ q' is necessary and neither 'p' nor 'q' is either necessary or self-contradictory." This theory, it seems to me, is subject to the disability I have mentioned. It is a familiar principle of axiomatic logic that if A is a thesis and if A ⊃ B is a thesis, then B is a thesis. As I understand him, Strawson would have us say that 'A is a thesis' means 'A is necessary' and that 'A ⊃ B is a thesis' means 'A entails B', and that therefore the principle in question is nonsense (which it is not). In OC, the principle is regarded as a conditional, necessarily true, about relations between certain logical formalities.

11. From *Principia Mathematica to *56*, p. 100. (Cambridge: Cambridge University Press, 1970)

12. The possibility of exemplifying logical formulas by expressions that are neither true nor false has spurred the development of many-valued logics. The plausibility of bivalence may be preserved, as in OC, by the insistence that interpretations of logical formulas be *statements*, expressions that succeed in identifying their subjects, and that are therefore capable of being either true or false.

13. Smiley, "Entailment and Deducibilty."

14. Anderson and Belnap, *Entailment*, p. 165.

Appendix

1. Were negative subjects permitted, the number of valid syllogisms would be greatly increased. The form a[–b(A)].–b[(A)]a[–c(a)] is inconsistent but it supports no syllogism because –b[c(A)] contains a negative condition.

Index